I0814451

STONE

STONE

ANCIENT CRAFT TO MODERN MASTERY

Richard Rhodes

PA PRESS

PRINCETON ARCHITECTURAL PRESS · NEW YORK

Dedicated to my teachers:

Signor Fabrini
Master mason, Siena, Italy

—

Oscar Costa
Quarry master, Vancouver, Canada

—

John C. Coldewey
Editor and mentor, Seattle, United States

Contents

Foreword / Paul Goldberger

I am writing these words while on a brief holiday in Rome, where stone, and magnificent architecture that has been made from it for more than two thousand years, abounds. Yet the most important part of Richard Rhodes's message in this book seems, in a way, almost to contradict the great architecture that surrounds me, because his point in writing is to make us realize that beautiful things made of stone are not only part of the past but also part of the present. This book is Rhodes's invitation to us to remember that stone lives not just in the great buildings we have inherited and that we continue to learn from; it lives equally, and perhaps more urgently, in the potential of our own time to make beautiful and meaningful architecture. Stone is not just the material of history, in other words. It can be a vital part of the present, and we need only the patience and the knowledge to understand how it might do so.

Making stone relevant in the present is Rhodes's life's work, and this book is, in part, the story of that life and that work. It is at once a memoir of his own experience with stone, an ode to the idea of stone, and a guide to the use of it. As a tech geek is to coding, Richard Rhodes is to the workings of stone, and sometimes he lets the geek side get the better of him, for instance, casually mentioning that "copings must be sloped a minimum of 5 percent so that water does not sit or pool on the capstone." Don't let a sentence like this distract you from Rhodes's passion, or from his story. When he talks about himself—which, being a modest man, he does less than he should in these pages—he inevitably has a wonderful tale to tell, not least when he relates his decision, after finishing graduate studies in London, to move to Siena and seek an apprenticeship with the ancient guild of stonemasons.

This was not something young American men were doing. It was not something very many young Italian men were doing, either, which is part of what intrigued Rhodes. The guild of Freemasons, founded in Siena in the thirteenth century, was then the oldest craft guild in Europe, and a living link to the builders of the continent's great cathedrals. It operated out of a medieval room near Siena's Duomo, and was more or less a secret society, passing down techniques from master to apprentice. Rhodes spoke no Italian (at least when he arrived), and he gives the clear impression that when he was invited to start working at seven o'clock the morning after his interview, it was not out of an eagerness to welcome this earnest young American into the secrets of the guild, but with the expectation that he would quickly be put in his place and slink back in shame to his home country. The Italian master masons viewed the United States as a siphoner of local talent, not a breeding ground for the next generation of stonemasons.

We know how this story ends, of course. The Italians underestimated Rhodes's seriousness and his natural affinity for working with stone, not to mention his tenacity. After his apprenticeship, which sounds something like a cross between Marine boot camp, a fraternity hazing, and a tutorial in technique, he became the first foreign national to be initiated into the Operative Guild of the Freemasons in its 726-year history.

In time, Rhodes returned to his hometown of Seattle and opened a business fabricating architectural stone, which he oversaw for a quarter of a century. He now works primarily as a sculptor in stone. He has fared better than the stonemasons' guild, which disintegrated in the 1990s, unable to survive the dearth of younger people wanting to learn the craft in the way that Rhodes did, and as other young men had done for more than seven centuries. The collapse of the apprenticeship system is a key part of Rhodes's story, and the real impetus for this book. He wants to record the lessons learned about how to work with stone, how to appreciate it, how to understand it—the unrecorded knowledge once passed from master to apprentice that is in imminent danger of disappearing.

It is common to think of the "art" of stonework as the act of sculpting it, and indeed it is, in a way, the highest art. But every single step leading up to that is also an art, from quarrying, to cutting, to transportation. Likewise, assembling its pieces into architectural form is an art. Rhodes's detailed review of the so-called Rules of Bondwork, the principles that master stonemasons pass to their apprentices, reveals this. The first rule, "Lay all stones in their natural bed," telling us that stone should be laid with its grain parallel to the earth, might seem obvious, but it is in fact enormously subtle, since grain itself is subtle, shifting, and not always clear, and the very act of quarrying involves knowing and understanding stone in the same way that a carver of wood must understand the material in the deepest sense.

Stone has a time—a time that moves slowly when it is in the earth, then rapidly after it is cut out of the earth. It needs to cure after quarrying, but wait too long, and it cannot be carved as easily or as successfully. Does any of this matter to those of us who will not be quarrying stone? Not as much as it matters to Richard Rhodes, obviously, but one learns much from this paean to stone, including why many works of stone construction fail—not only celebrated collapses like the Beauvais Cathedral, but also more mundane failures like Empire State Plaza at Albany, where the walls' structure is strong, but the stone veneer was installed incorrectly and failed. Modern architecture has witnessed more than its share of such failures. "Know thy stone," Rhodes is telling us.

Rhodes is realistic about the world. He knows that we no longer build out of great blocks of stone like the ancient Greeks or Cambodians, and that most of the stone used in building today is not there to hold up a building but serves as a veneer, a beautiful cover over a different structural system hidden beneath, not unlike the glass curtain walls applied over the steel frames of modern skyscrapers. Rhodes does not wish for architecture to return to the fifteenth century, or even to the nineteenth. But he does want us to regard stone with greater understanding, with a deeper feel for its qualities as a material, and he is candid about the risks of not doing so, and about how facile and cheap so much contemporary stonework is. "We have lost, in fact, stone's connection to the sublime," he writes, and that is one of his fundamental points. At a telling moment, he asks, "How did previous civilizations create work of transcendence? With such primitive technology and so little precedent, how did they manage to make work still capable of stirring us emotionally? And how will we charge our modern stonework to speak with equal eloquence of our lives and time?"

Upon finishing Rhodes's manuscript I took a walk, first to the Pantheon, then on to Gian Lorenzo Bernini's great Sant'Andrea al Quirinale, Francesco Borromini's even greater (to me) San Carlo alle Quattro Fontane, the Palazzo Barberini, and finally the Piazza Navona. Everywhere there was stone, and every stone in every one of the buildings, every stone in every piazza, was different. Each individual piece had been shaped—as carefully wrought a part of its respective story as every word, every verse, of Shakespeare, or every phrase of Beethoven. Together the stones made works of art, and together the works of art made the city, and the civilization. None of this would have happened, Rhodes wants us to know, if the builders had not understood their material. The physical world has its rules, and in the making of things of stone, it is essential to know them. "A central truth," he tells us, is that "aesthetic triumph arises directly from innate structural success."

And yet for all that Rhodes recognizes and celebrates about the realities of stone and its function in the physical world, his true agenda is to show us something else. He wants us to feel that the greatest gift stone can offer is to transcend the physical world and give us a whole other level of experience. He knows that architecture provides not just shelter but emotion, and that stone that has been crafted by hand embodies emotion profoundly. Indeed, it can bring us to the sublime. How to achieve that in an age that has no patience for apprenticeship, that operates by technology that makes things ever thinner, ever lighter, ever less possessed of natural gravitas? That is the challenge Richard Rhodes sets before us in this book.

Since earliest times, stone and stone architecture have provided canvases for human expression. Petroglyphs often predate civilization as we understand it and speak to the universal urge to communicate to future generations. Moab, Utah

Introduction

> **Not least these timeless stones should cause us to question the real nature of permanence and the real nature of what it means to be human.**
>
> —KEITH CRITCHLOW[1]

Making, building, creating something of lasting beauty is an essential part of being human. It is also one of the crucial characteristics of our existence that separates us from most of the animal kingdom. The impulse may be tied intrinsically to a fundamental need to communicate our legacy to future generations: a way of speaking of our time and our concerns.

Prior to the twentieth century and the mass deployment of concrete, the majority of humankind's abiding contribution to the built environment was carved or built with stone. Stone's indelible qualities have conveyed every chisel stroke forward, allowing us to read its expression today, tomorrow, and for eons to come. As the photographs in this book aim to convey, this expression remains legible, wide-ranging, and vitally varied—as multifaceted as humankind itself. No wonder so many civilizations have turned to stone for both building and creative expression. Being quintessentially of the earth, stone remains fundamental to our sense of home and our place in the cosmos. Indeed, our relationship with stone is so deep and intimate that we named the beginning of human history the Stone Age.[2] Older than life itself, stone is our anchor, its timeline bending both forward and back.

When we discover rock in the landscape, or quarry it from beneath the earth's surface, it contains only potential, nothing more. Our very language reflects this essential understanding. When we say "on the rocks," we mean "quite destitute of means." Rock becomes stone once we work it, once our labor has carved or shaped it for our use. This transformation is uniquely human: Man the Toolmaker, knapping flint weapons and shaping stones for specific use and unique purpose. To this day, our Stone Age impulse remains and drives our connection to the material. Through human hands, stone's visual range becomes infinitely varied and expressive. Its hard but ultimately pliant surface captures our intention, holds it, and carries it firmly to the future. Wrought stone represents a summation of sorts, the human effort of evolved tooling, spatial cognition, and practiced handcraft. A great intensity—aligned with determined purpose and plan—working toward a singular objective. Stone captures this effort like no other material, preserving and communicating a touch of the soul of the artisan who crafted it. Stone's final aesthetic derives from both the act of crafting and the craftsperson wielding the tools.

A peculiar kinship exists between humans and stone. Beyond material comfort, the subtle wearing down of stone records the passage of human and geological time. It records our human activity, whether from carts traveling the ancient roads of Rome or foot traffic over the centuries, becoming an energized artifact of civic life. Stone also captures the history of the craftspeople who shape it, forever recording their labor in the way it receives carving and tool marks. Stone's receptive quality allows the natural vocabulary of the material to expand into a textural language. Even with the simplest of hand tools, human beings have created new stone textures that catch light, cast shadows, and form space. Stone's innate permanence obliges our efforts, preserving human expression for eternity.

I start this book by looking at those very hand tools and the small technological advances that have prompted shifts in stone expression over the millennia. These shifts occurred unevenly. Some civilizations created deeply expressive works with the most basic of tools, while some of the most technologically advanced cultures in human history floundered in their attempts to create a legacy in stone architecture. By exploring the work of ancient peoples who contributed to our shared architectural legacy, we build context for the deeper questions that the material inspires: the yearning for memorial and the expression of what it means to be human.

Stonemasonry

Starting in earliest times, the long story of our world's architectural history has been the history of a single craft: masonry. When we study stone, we study ourselves and our place in the world. We confront human endeavor—its achievements and failures, dreams and aspirations. The people included and excluded by walls we have built, the communities joined by the bridges we constructed over rivers and chasms, the values and honors memorialized by our monuments, are all captured in stone's lithic mass.

Interestingly, new modes of expression do not happen in a vacuum but align with a culture having something urgent to say. Moments of high expressiveness may coalesce due to profound religious alignment, such as that which gave rise to the European cathedrals or, as in the case of the last half century in the United States, cultural shifts of utopian idealism which sparked expansion in all arts simultaneously. Such moments may be hard to identify from within. But with the benefit of hindsight, it becomes obvious that humankind has turned to stone to record its myriad transitions. In this way, stone remains more than a building material, since it arguably carries the primal emotional legacy of the estimated five hundred generations of human habitation that came before us.[3] Much of stone's beauty arises from the emotional response awakened by that ancient legacy. When we walk the streets of Europe, Africa, Latin America, or Asia, the worn pavements stir us, arising from that shared human connection to those who have come before. The simple act of running our hands along a worn stone railing as we pass connects us irrevocably to our deeper shared history. Stone is literally one of the touchstones to our world. We experience stone architecture by physically and psychically interacting with it.

As will become clear in the pages ahead, many of the best past expressions of stone follow a traditional rule set, but that rule set hardly tells the whole story. How, why, and when traditions evolve are some of the most compelling questions we need to consider.

A Personal Encounter with the Sacred Rules

My own introduction to stone was not glamorous. I had finished my graduate studies in London and was making my way to Siena, Italy, possessed by exceptionally vague notions on how to relate my studies in medieval drama to Europe's oldest guild. The guild was called the Freemasons, and it had been founded in Siena in the middle of the thirteenth century. These practitioners were not the secret-handshake Masons of the benevolent societies found in the United States, but rather the original guildsmen of mason-architects, the professional descendants and heirs of the cathedral builders.[4] From my research in medieval and Renaissance theater, I had discovered that the Freemasons presented a living connection to medieval times and the creative spirit that had built the masterworks of Europe. I was deeply intrigued.

Not unlike many young people taking their first tentative career steps, I was long on ambition and short on practical ideas. I spent my entire "savings" while biking from London to the ancient Roman city, didn't speak Italian, didn't have an introduction, and didn't know anyone in the entire country. What could possibly go wrong?

Despite all this, through a series of inexplicably synchronistic events, I was soon climbing worn stone stairs near Siena's marble Duomo into a vaulted space that was the guild's medieval lodge. Thick, leaded windows formed a clerestory, dimly illuminating the faded tapestries lining the walls. The air was heavy and humid with what I would come to recognize as the moisture that masonry buildings wick upward from the earth on which they rest. The room was large, and lined with straight-backed chairs—chairs so heavy, I imagined four men being needed to drag just one into position. This lent the space a formality I had never experienced on the US West Coast, where I'd grown up. Standing immobile in the center of the room stood Signor Fabrini, the master mason of the famed Sienese guild. He was not tall, slightly hunched in fact, yet radiated an unmistakable gravitas.

Approaching him slowly, I bowed ever so slightly and waited. Unsmiling, he eyed me in silence. I felt intensely uncomfortable and began to explain that despite my having zero experience, he should hire me. With the hubris of youth I continued, "What I really want is to apprentice with the ancient guild." Alas, my carefully memorized explanation in Italian proved unintelligible, a calamitous collision of mispronunciation and terrible grammar. I rattled on, and he just looked at me with his soulful, tired, unblinking eyes. Determined to hold my ground, I locked on to his gaze. This wasn't an act of defiance; it's just that time had stopped entirely in that long, frozen moment, and blinking never occurred to me. Looking back, many things had not occurred to me. I hadn't formulated a plan B. This was my shot. Finally, after winter and summer had seemingly come and gone, he said slowly, "Domani, alle sette" (tomorrow at seven), turned, and left. "Well, that was hard," I thought as I made my way back to the street.

I had no idea.

The next morning, I arrived early at the jobsite in the center of town, taking the bus from the farmhouse squat I had found on the road to Grosseto. To my surprise, Signor Fabrini was already there, supervising the delivery of the day's materials: a cubic meter of sand, three bags of lime, and seven bags of Portland cement. This was the early 1980s, when a bag of Portland still weighed 50 kilograms, or 110 pounds.[5] He pointed, and I grabbed the primitive shovel. I couldn't help but smile as I noticed that its rough handle was made from a tree bough pinned primitively to a hand-forged metal blade. It looked positively medieval. "You're not in Kansas anymore," I thought to myself as I jumped in to pitch sand out of the narrow pedestrian street into the entryway of the medieval building. The project at hand was to convert the structure's top two floors into a small hotel.[6] As the masons arrived, without prompting, each grabbed a bag of cement, deftly tossing it on their shoulders. Snagging the leather lunch satchel they had momentarily set down with their free hand, they began the slow climb to the fifth floor, where the construction work was being staged. Seeing the advanced age of this group (it turned out, all the masons were over sixty) and feeling in the best physical shape of my life, I confidently hoisted the last bag of Portland and took up the rear on the slow climb up the internal stair.

It is hard to describe just how exhilarated I felt climbing the worn brick stair that would come to define my initial months of work with the guild. I had never done real physical labor, let alone had a chance to work shoulder to shoulder with the heirs to the cathedral builders of Europe. I might have been carrying bags of cement, but this was heady stuff.

I was only forty-four steps up the staircase when I felt the first bolt of real panic. It arrived as a cold sweat, seeming to burst through my skin, the physical alarm arriving long before conscious awareness. In an instant, my euphoria drained away, and all that remained was my thundering heartbeat, as loud as I'd ever heard it, ringing in my ears. "How many flights did he say?" I thought to myself. Perhaps I hadn't understood. "Surely these old guys aren't going to carry these heavy bags up three more flights?"

When I reached the work's staging area on the fifth floor, the pounding pulse had migrated to behind my eye sockets. I couldn't see clearly for the intense throbbing. I was choking back my breakfast, hoping to keep it down. The others calmly set down their bags, looked at me, and began to laugh uproariously. My bright-red face said all they needed to know.

The rest of the day was "easier," at least on my ego. They explained that all I needed to do was bring the daily cubic meter of sand (3,650 pounds) up the 110 steps. If I really hustled and got done early, I would still be paid for the entire day. They motioned to some old rubber buckets, and I got to work. Walk down five flights, fill two buckets, carry them up five flights, empty buckets. Simple. Repeat this round trip fifty times a day, each time carrying roughly seventy pounds, for a total of 5,610 vertical stair steps. As grueling as the work was, I felt strangely uplifted. The first day took me "only" thirteen hours, or fifteen minutes per round trip.[7] A "hard day" in my previous life was a twelve-hour stretch in the library stacks reading Old English. Counterintuitively, this work seemed easier. I found it strangely meditative. I quickly learned one of the secrets the older masons demonstrated while carrying the heavy cement bags up the stairs. If you move slowly, even under great strain, and keep your heartbeat just under the lactic threshold, you can go all day.[8] As they cautioned me that first day, "Riccardo, piano, piano, ha fatto il duomo di Milano" (Richard, slowly, slowly, built the great Cathedral of Milan).

As it turned out, this trial, or *la prova*, as it is known in the guild, stretched out for four months. I used the time productively, gathering snippets of the local Tuscan dialect with each emptying of the buckets, working to decode the new words on my way down the stairs. But physically the work was brutal. I lost a lot of weight. I found out many months later that the severity of *la prova* was to get me to leave. "Always better to find out what someone is made of up front," as they would say.

The guild had lost generations of masons to the United States. I was the first to travel the other direction and appear at their door. For them it was all deeply suspicious. Who knew the real intentions of this left-handed foreigner? The left hand is *la sinistra* in Italian. In the end, they finally initiated my apprenticeship, but only after I'd lost so much weight from the work that they'd grown worried about my physical condition. As the months wore on and the extreme exertion was obviously affecting my health, I was finally encouraged to begin practice with stonemasonry's emblematic tools, the hammer and chisel. Who could have imagined that from such an inauspicious start, I would become the first foreign national to be formally initiated into the Operative Guild of the Freemasons in its 726-year history? In subsequent chapters, I frequently return to these formative years to illuminate key concepts that were revealed to me during this time.

Upon completion of my apprenticeship in Siena, I worked hands-on with the tools of masonry construction and stone design for eleven years in my first bootstrapped installation company. While honing my craft, I continued my professional studies by apprenticing with a master quarryman and stone splitter from the Azores, our shared communication a patois of Portuguese, Italian, and Canadian French slang. For ten years, I manufactured hand-carved architectural stone in Seattle before moving my operations overseas. That company grew quickly and eventually involved stone production and quarry operations that employed architects, engineers, and construction managers in a global operation that ran seven days a week in twelve time zones. At its peak, the company shipped a container of handcrafted dimensional stone somewhere in the world every day.

After twenty-five years and dozens of private commissions and public artworks completed, in 2012 I turned my attention to sculpture full time. Stone sculpture inhabits a realm of material nuance, detail, and precision beyond that of ordinary masonry. It requires time not ordinarily available in the midst of international commerce. Most notably, the practice of sculpture has allowed me to explore my original impulses encountered during my masonry apprenticeship, extending my long journey of discovery in the medium. Those nascent concerns will be traced through the chapters of this book. The "soul of the artisan," the "legacy promise," and the "expression" of stone remain themes that resonate both with my sculpture practice and with the very core of my being.

My exploration of natural building stone and its crafted expression is inherently personal. It is informed by decades of hands-on experience, and in the pages ahead, I include stories from my journey that have supported my knowledge and conclusions. It is important to point out up front that I am a nonacademic scholar. The scholarship I cite is always from others, as detailed in the endnotes. My unique contribution comes from my years with the tools and "felt knowledge"—the sense extracted from the stone itself.[9]

An important additional aspect of what I want to share is the body of knowledge I encountered while apprenticing in Siena, within the guild itself. Although there will always be those who try to cut the line and skip the paying of dues, the fact remains that acquiring traditional knowledge and skills takes years of watching and imitating, long consideration of examples given, and the hard-won experience of personal attempt and failure. With that in mind, it may come as no surprise that given our shortened modern attention spans, the guild structure became unsustainable and has now collapsed almost entirely. Indeed, the last master masons have for the most part disappeared. My own guild in Siena died out in the early 1990s after more than seven hundred years of unbroken tradition. There were simply no more young people willing to put in the years to learn the trade and carry it forward. The same apprenticeship system that formed a continuous chain of knowledge has likewise vanished altogether in the United States.[10]

Along with this disappearing community of workers have gone the ratios and formulas, the design principles of patternmaking and sacred proportion. These are the ancient secrets of Freemasonry known as the *Sacred Rules of Bondwork* and the *Sacred Geometries*. Unbeknownst to me at the time, the elderly masons I worked with were some of the last practitioners of the guild of Freemasons, the last formal inheritors of the ancient design principles. The Sacred Rules, as they are collectively known, were considered the guild's most important and closely kept property.[11] In our secular culture, the word "sacred" can confound or turn many of us away from looking more deeply at the concept presented. So, going forward, I will refer to this once "sacred" rule set as the Rules of Bondwork.

Historically speaking, from a purely business perspective, it was not in a guild's interest to share knowledge with the uninitiated. As inheritors of ancient mathematical truths and chemical processes, Freemasons saw themselves as the keepers of a secular faith. The stone design rules remained the private source of their power and competitive advantage, and their livelihoods depended upon their trade secrets.[12] As the last apprentice of the guild (as I came to be known), it has been hard to witness this tragic decline. The Rules of Bondwork have never been set down fully, let alone explained and illustrated. In the last decades of my career, it arises as a supranatural duty to share the ancient knowledge before it is forever lost. This book is my answer to that call, sharing these secrets for the first time.

The Rules of Bondwork offer us a unique view, a frame through which it is possible to decode the last five thousand years of stone architecture and expression. Even when they could not possibly have encountered the ancient rule set, we can see cultures working it out and often coming to the same essential conclusions.

One unexpected result of the collapse of the masonry guilds has been that my research on stonemasonry in published sources has been exceptionally difficult. The body of knowledge was held by manual tradespeople. Additionally, the masonry profession has been mostly ignored by professional scholars. Indeed, an alarming amount of what I found in published work is seriously flawed, written by authors whose hands never touched stone-working tools and innocent of the most basic knowledge any hands-on practitioner would know. Consider the eminent Egyptologist I. E. S. Edwards, whose otherwise exceptional book *The Pyramids of Egypt* explains, "Since copper is the only metal for tools known to have been available in Egypt before the Middle Kingdom, it has been supposed that the Egyptians had mastered a process, now lost, of giving copper a very high temper."[13] Tempering metal makes it harder. We can temper by hammering, or by heating and rapid cooling, a process called quenching. Despite Edwards's magical thinking, copper tempers minimally by hammering and not at all by quenching.[14] It can be made harder only by mixing it with other metals (such as tin, to make bronze), knowledge that, for the ancient Egyptians, lay a millennium in the future. In fact, Egyptians in the twenty-sixth century BCE figured out that the sharp and hard quartz dust of granite will embed into the head of a shaft of soft copper, making a serviceable, if primitive, chisel. It is not a terrific tool by any stretch of the imagination, but it will slowly cut stone.

Most of the expertise explained in the coming chapters has been acquired in the same way that masons have always learned: from actual practice, and with the guidance of an oral tradition passed from father to son, master to apprentice.[15] While I first encountered the Rules of Bondwork in Siena, other aspects of this knowledge were received in quiet conversations with elder masons as they faced the end of their working lives. In the decades since my formative apprenticeship, masters from all over the world have shared their techniques with me, and I have tried to bring their insights to this text wherever possible. All of this knowledge has been hard-won, for only when trust is established and credentials vetted do real exchanges begin. I am not an academic or architect by training; I am a stonemason. I came up through the ranks by following what felt right and natural to me and by learning from the masters. As a reflective practitioner, I lecture widely, write papers, and have been preparing this book for more than a decade.

In my own life, the Rules of Bondwork provided the original spark that ignited my interest in stone and turned my life force toward the material as an all-encompassing vocation. They allowed me to decode the built environment and express my own synthesis of our cultural moment through my practice. What a great gift it has been to see these invisible principles made visible in expression through stone. This book shares my appreciation of the astonishing collection of masonry buildings that form our shared global legacy.

The surviving architecture of stone offers us evidence of the very best and worst of humankind. The ruins of stone and masonry architecture testify to war and destruction, to the rise and fall of cities and civilizations. Our heroes and villains traveled stone roads, trod stone battlements. We find stone harbor towers to deter pirates, stone monasteries to shelter saints, giant stone walls partitioning continents. Stonework offers us a palimpsest of our cultural history. In this way, it carries its creation myth on its sleeve, and even a layperson's understanding of archaeology can trace historical narratives.

Granite weathers one inch every ten thousand years.[16] When we build and design with stone, we make an enduring statement about ourselves, our skills, and our times. Stone obligingly carries that statement forward for future generations and cultures to read. And, as I have learned, work we create in stone will very likely outlive us, our family name, our country, and our civilization. Knowing and embracing that seminal fact should lead us to think carefully and well about the outcome of our work. If we use this material merely for a "twenty-year building life"—the common standard for most US construction today—we forget the privilege of working in stone and we neglect its power.[17]

Whether applied as cladding to the world's tallest buildings or glued into private jets, not all our contemporary applications are well considered or thoughtfully installed. My impression is that designers, masons, and architects working with the short view err in ignorance, not intention. After all, no one sets out to do poor work, although working on the cheap sometimes comes into play. The larger issue is one of education and guidance. Never has stone been more plentiful, yet never has the knowledge of how to build with it been so elusive. My own quest for serious written work about stone and masonry bears this out. Searching the literature, I found that virtually all of the important books on these subjects have been out of print for more than seventy-five years, or are not available in English. Worse, many of the books and articles I did find were a jumble of misused and/or badly coined terms. Proprietary language and invented terms are routinely employed to obfuscate the origin of the material or surface treatment rather than to illuminate its purpose and use. Stone installers and suppliers, usually self-taught, may make up their own terms to describe their finishes or patterns. With no standard language, no accurate and consistent vocabulary, those who want to use stone and those who provide it have arrived at an impasse. Given this state of affairs, is it any wonder that even design professionals who have worked for years with stone, otherwise confident in their technical vocabularies, are often unable to clearly communicate their design goals for masonry?

You will find my research-validated terminology for stone in the glossary following the text.

The pair of epic sculptures known as the Colossi of Memnon were carved from a single megalith in 1350 BCE. Each one stands 60 feet in height and weighs an estimated 720 tons. Unlike the Sphinx at Giza, which was reportedly used for target practice by Napoleon's army, the Colossi were likely damaged in a massive earthquake that destroyed the temple they once announced. Luxor, Egypt

The defensive walls around the city of Jerusalem are pocked with bullet holes principally from wars in the last seventy years. Jerusalem, Israel

TOP
Since earliest times, societies have reflexively turned to stone to memorialize the loved and the admired. In this way, stone, whose age we now measure in the billions of years, has acted as a mythopoetic portal into the deep time of the cosmos. The nearly universal act of inscribing names into the lithic medium speaks to the deep impulse to align our brief lives with eternity. Boston, Massachusetts

BOTTOM
A stele is a rectangular monolith or slab with a rounded top. While this type of stone tablet frequently memorializes fame or victory, it has just as often recorded mundane civic matters, for instance, tax and frontier boundaries or geographic distances, as in milestone markers. Jinan, China

TOP
Contrary to popular opinion, the Sahara Desert is not an endless sea of sand. Rather, it contains fragments of limestone whose surfaces are beautifully honed by the natural friction of sand and wind. Finding limestone hundreds or thousands of miles from the nearest ocean informs our knowledge of the age of the Earth and our place in the cosmic timeline. Erfoud, Morocco

BOTTOM
Some three hundred years before Western Europeans arrived in the Americas, magnificent stonework was undertaken by the Anasazi, the ancestral Pueblo Indians in the US Southwest. Alas, like their very culture, none of their stoneworking terms have survived and entered the Western stone vocabulary. However, their stone structures carry on. Hovenweep, Utah

The dry climate of the US Southwest allowed wood beams and lintels to be incorporated directly into the stonework without rotting. Although the pattern bonding provided little strength, the stones are set with great intention throughout the wall section, as seen in this exposed wall end. Their well-bonded interiors ensured that the walls survived for a millennium, and they remained the largest buildings in North America until the nineteenth century. Chaco Canyon, New Mexico

This Book: Summons to Foundational Knowledge

Following the opening two chapters, which give a cultural and technological context for architectural expression in stone, my explanation of the Rules of Bondwork occupies the remaining chapters except the last. Once the Rules have been explained, we turn to the way stone is used today. The book ends with a manifesto, a passionate call for a reconsideration of the way that stone and our other precious resources are deployed. It is a strident call to action to motivate the design and stone community to urgent change.

Today, the push for construction velocity, combined with dubious technical "advances," has allowed carelessness and poor judgment free rein. Contemporary stonework is too often deployed without thought or skill. Paradoxically, as material engineering and construction techniques have advanced, we have lost sight of some basic, inherent qualities of the material itself. Humankind's hundreds of generations of experience and dedication to stone has produced a deep understanding of its nature, and yet we have lost familiarity with the expressive character of stone, and ultimately with its poetry. We have lost, in fact, stone's connection to the sublime.

The reality of this loss drives my mission for the book. It prompts me to address some of the critical questions too often ignored in the contemporary practice of architecture and design. How did previous civilizations create work of transcendence? With such primitive technology and so little precedent, how did they manage to make work still capable of stirring us emotionally? And how will we charge our modern stonework to speak with equal eloquence of our lives and time?

I argue that only through the study of the ancient Rules that governed stone's past use can we position ourselves to interpret the material in fresh ways, and then move toward a richer and more informed modern architecture. Learning the hard-won Rules and serious prohibitions of stone's past use helps us appreciate our common environment, and points the way toward more natural design and its application in the present. Importantly, knowing stone's historical use does not condemn us to work in obsolete codes or classical vernaculars. Rather, knowing the Rules allows us to choose to work within their strictures, or break them deliberately for informed artistic expression. After explaining fundamental concepts related to the design and application of stone, this final chapter is my attempt to align our future buildings with the five-thousand-year history of architectural stonework.

It is widely acknowledged, if only vaguely understood, that astonishing historical structures were built in keeping with the Rules of Bondwork. How to align current and future stone use with the traditions of the past is a matter much less explored. To that end, this book aspires to be a beginning—another stone in a wall made of written, rather than oral, knowledge. Perhaps it will inspire other stone masters to step forward and reveal parallel traditions handed down to them. Modern science and management principles have taught us that the open sharing of knowledge is a powerful catalyst for the advancement of society and organizations. Knowledge is one of the few things that can be given away without losing it. Yet the old story is still the true one: mastery requires decades of practice and thoughtful execution.

I.

A SHORT HISTORY OF STONE USE AROUND THE WORLD

Chapter One

TOOLS OF TRANS-FORMATION

Art comprises the whole of man's works, the material outcome of his thought as expressed by his hands and through the tools of his invention.
—JOHN HARVEY[1]

One way to explore the use of stone through history is to follow the development of the tooling used to craft it. Each small technological advance allowed for an expansion of stone's visual range. Such shifts occurred unevenly, however, since some cultures, like the Egyptian, had both adequate tooling and advanced mathematical concepts, while others, like the Inca, were limited in their technical and mathematical facilities yet still driven to express something urgent within their souls.

In general, the tooling technology of stonework is small. There are a finite number of efficient tool shapes for working stone, and stoneworkers in unconnected civilizations came to remarkably similar conclusions concerning what worked best.[2] In the earliest civilizations, the tooling centered on percussion and abrasion. Consider the miraculous Inca stonework begun at the middle of the fifteenth century, created by a culture with neither access to metal tooling nor even the discovery of the wheel. In fact, the Inca did not even have a written language, but kept track of tonnage and commerce by tying elaborate knots called *quipu* on string cords.[3] Yet in an eighty-year burst of communal artistic exuberance, this Stone Age culture assembled some of the most compositionally and technically challenging stonework ever seen.[4]

One small part of the Inca success story literally fell from the sky when a large meteorite struck the earth nearly four hundred miles from the Inca imperial fortress, Sacsayhuamán, overlooking modern-day Cuzco. Legend has it that the king ordered the gathering of the meteorite fragments for use as hammerstones. They now possessed a primitive tool, forged in the outer atmosphere and having the hardness of carbide steel, measuring 9 on the Mohs hardness scale (which ranges from talc, 1, to diamond, 10).[5] Hammerstones are a percussive tool; one beats the tool against the stone, incrementally crushing the latter's surface to shape it. The harder the hammerstone (or the softer the stone being shaped), the easier the work. Of course, in the case of the Inca, thousands of hammerstones were ultimately required for construction, an amount far exceeding the abundance of meteorites collected, so the stonemasons used a wide variety of other hard stones, including common river cobbles,

quartzite, and chert.[6] The archaeologist Jean-Pierre Protzen alone found sixty-eight hammerstones in a single Inca quarry called the Llama Pit: "The weights range from a couple of hundred grams to hammers of 8 kg. [17.5 lbs.], with two groups in between, in the 2–3 kg. [4.5–6.5 lbs.] and the 4–5 kg. [9–11 lbs.] ranges."[7] All the hammers Protzen found had a hardness of at least 5.5 on the Mohs scale. Although this hardness was comparable to the andesite stone the Inca were crafting, the andesite was brittle and shattered on impact.

Protzen also points us to Tiahuanaco, near the southern shores of Lake Titicaca in modern-day Bolivia, whose residents achieved similar technical results, likely without the cosmic carbide of the Inca. The Tiahuanaco culture reportedly started in 200 BCE and died out almost four hundred years before the rise of the Inca civilization.[8] Little is known about it, but its stonework shares much of the ethos of the later Inca. Using hammerstones to shape the raw material, they managed to produce cut stone construction of "unsurpassed maturity and perfection."[9]

Technological advances expanded the range of expression and allowed architectural shifts in the way stone was used. Transitioning from hammerstones to copper and bronze tooling, for example, unleashed a torrent of creative change. Without metal tools, the shaping of stone had been limited to more roughly polygonal or amorphous shapes. Although it is technically possible to create sharp arrises (edges) with hammerstones, only metal tools—even soft copper or early bronze ones—make it repeatable at scale. The action of hammerstones—the collision of two stones of different hardness—allows only for crushing of stone rather than the chipping or flaking possible with metal tools. Metal tools can both chip and crush, and a good stonemason selects from their various hand tools for different effects. Because the finer action of stone carving is largely a chipping process, it is not surprising that the very earliest advanced cultures with access to metal tools incorporated carved elements into their stone architecture.

The addition of tin to copper to create the alloy bronze flourished for almost one thousand years before the Egyptians exploited the technology. Discovered about 3000 BCE in the eastern mountain areas of Syria and western Turkey, where both copper and tin are found in abundance, bronze was widely used for weapons and tools in the wealthy culture of Mesopotamia far to Egypt's northeast. The British archaeologist Henry Hodges

OPPOSITE
The Inca created precisely fitted walls from massive stone blocks. The largest of these, pictured here, measures approximately 13 feet in height from current grade and is estimated to weigh 350 tons. Cuzco, Peru

TOP
The Egyptians used hammerstones to quarry and carve their granite obelisks starting 4,500 years ago. But it was their early copper tools that enabled the carving of the elaborate hieroglyphics. Of the hundreds of obelisks that once adorned the temples and palaces of Egypt, only forty have survived, "all of these more or less mutilated, only nine now standing in Egypt, ten fallen and broken, and the greater part carried away to foreign lands." [Alexis A. Julien, "The Misfortunes of an Obelisk," Journal of the American Geographical Society of New York 25, no. 1 (1893): 89.] Luxor, Egypt

BOTTOM
This ancient Egyptian carving is all the more impressive considering the soft copper tools that made it. In the bas-relief method, stone is carved away such that the imagery is raised in a shallow plane. All of the sculptural nuance must be accomplished between the two planes that define the start and the final depth. Luxor, Egypt

writes that although Egypt was "endowed with large quantities of copper ore, the country was utterly devoid of tinstone."[10] Still, using only soft copper, the Egyptians managed to create hand tools strong enough to carve crisp dedications into their granite obelisks, itself a daunting accomplishment. In fact, the quality of their expression seems to have been largely unaffected by the arrival of bronze tooling in about 2000 BCE. Apparently they were already producing what they desired, and although the harder bronze may have sped up the work, it did not materially change its expression.

This drawing on the wall of the tomb of Rekhmire at Thebes clearly shows early chisels at work in the creation of ashlar masonry (finely wrought rectangular blocks with small joints). Judging from the scale of the blocks, it likely illustrates the construction of the Pyramid of Menkaure at Giza. Luxor, Egypt

Curiously, the transition from the Bronze Age to the Iron Age was not particularly important for building with stone. The first steps of tool hardening seem to have helped productivity, but did not produce significant new expression. Indeed, Greece and Rome produced little expressive work with their early iron technology that the Egyptians had not already accomplished with hammerstones and copper chisels. Despite being credited with inventing the tooth chisel (a chisel blade with grooves like teeth cut into it) in the sixth century BCE, Greek masons did not initially surpass the exceptional work of the Egyptians, which had been carved in much harder and more difficult material with more primitive tools.[11]

Yet this is not the full story. Even incrementally stronger metal tools allow for smaller pieces to be worked; essentially, the impact and percussion of the metal chisel are more locally contained than when using blunter tooling. A hammerstone sends destructive vibration and pressure through the material, and the percussive blows can break the stone apart altogether if it is too thin. To resist this vibration, the receiving stone must be of a scale large enough to absorb the impact without cracking structurally, whereas a simple point chisel of marginal hardness successfully localizes the impact of the blow, allowing the size of the piece being worked to shrink dramatically. This begins to change a fundamental equation in stone sizing and allows for more delicate work to be accomplished in significantly smaller pieces. For example, in figurative Egyptian sculpture, the carved arms are typically depicted as crossed against the chest or flat against the body, such that the arms, hands, and fingers are supported by the larger mass of the figure itself. However, the small improvements in Greek and Roman metallurgy allowed for the arms (and sometimes even fingers) of sculpture to extend into free space, unsupported by the mass of the torso. Almost all of this work happened in relatively soft marbles,

One characteristic of Egyptian sculpture is that the figures' extremities remain connected to the larger mass, an artifact of the use of hammerstones to accomplish the carving. More delicate forms might shatter from the percussive blows. Luxor, Egypt

LEFT
The earliest process for creating stone slabs from quarried blocks imitated the sawing of wood. This method worked particularly well on soft limestone (marble).

RIGHT
The earliest stone shelters were made from slabs or sheets broken off naturally formed outcrops. As more sophisticated tooling developed, stone was quarried and then chiseled or sawn to the desired shapes. Qufu, China

where delicate shapes may also be achieved through simple hand grinding of one abrasive stone against another, softer, material. Still, the work of roughing out the material into the required form was initiated with a metal point chisel, localizing the impact of the blows against the stone of a more delicate form being shaped.

Early Mechanization

Creating relatively thin slabs of stone is technically difficult. The Egyptians successfully used handsaws with copper blades fed with sand slurry to slowly wear away even the hardest of stones,[12] and handsaws were used extensively by the Greeks and Romans on much softer materials like tufa and marble. From the perspective of expressive change, the real breakthrough came when blades were connected to nonhuman power, effectively creating one of the first stone-working machines, enabling the cutting of thin slabs of stone of a consistent thickness without direct human exertion.

Gang saws are large, room-size stone-cutting machines with flat blades that are assembled, or "ganged," together in series. Like the handsaws that preceded them, the blades are of mild steel and not sharp. Rather, these saws cut by abrasion, wearing the stone down. A popular variation developed in the 1920s used braided steel cable instead of steel blades, but the principle remains the same. In its most basic iteration, water (or, later, steam) is the power source, driving the blades forward and back. Water is flooded across the blades, carrying a slurry of sand or steel shot that is ground into the stone to create the abrasion for cutting—coarse sand for hard stone, and fine sand for softer material.[13] The water cools the blades or cable and helps the steel retain what little temper it has. The action of the gang saw effectively wears away the stone, creating slabs of consistent thickness.

In fact, the technology of the gang saw is Greek, with water-driven saws working their way through stone blocks as early as the third century BCE. This remarkable feat required the conversion of the rotary action of a mill to the linear action of a saw by way of a primitive crankshaft. An extraordinary diagram of this

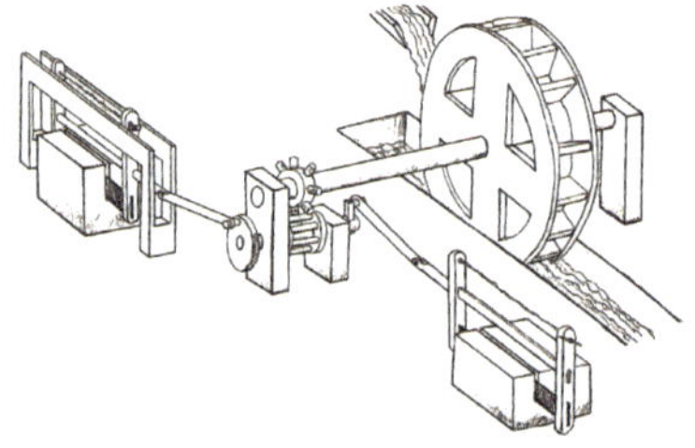

ABOVE
The depicted mechanism for converting the rotary action of the mill to the linear action of the stone saw is carved in bas-relief on the sarcophagus of a wealthy Roman miller who lived in the third century CE. Known as the Hierapolis saw-mill, the machine was the first to in… crankshaft and …erapolis, Turkey

…gang saw …labs of …e same time, …nt that …as applied

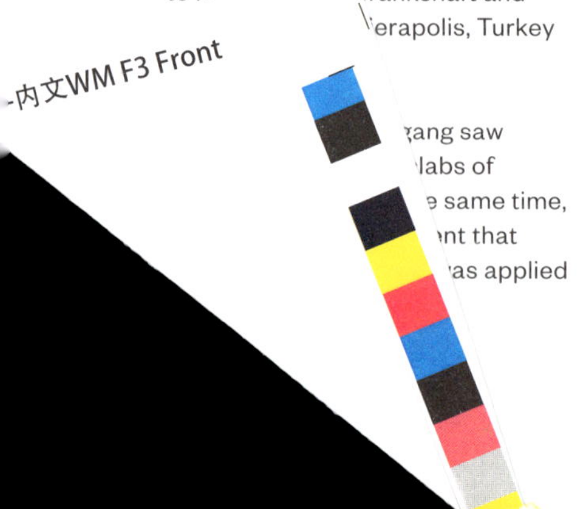

mechanism was carved on the sarcophagus of a wealthy Roman miller named Marcus Aurelius Ammianos at Hierapolis (in modern Turkey).[14] There are other good examples in the ancient ruins at Ephesus, Turkey, and Jerash (Jordan). Additionally, we learn from the poet Decimus Magnus Ausonius (ca. 310–395 CE) of the river-driven sawmills that supplied the stone slabs for the imperial city of Treves. Ausonius reports that water from the swift rivers Celbis and Erubris, "famed for marble" and "for glorious fish...turns millstones in furious revolutions and drives the shrieking saws through smooth blocks of marble, hear[d] from either bank [as] a ceaseless din."[15]

Harder Metal, Expanded Visual Range

Between the fall of Rome in the fifth century and the rise of the European city, generally viewed as commencing in the tenth century, masonry structures were built primarily with shelter and safety in mind. Protecting the people inside, keeping out the wind and rain, and holding up the roof were reasons enough to turn to stone. And of course stone is fireproof, and accidental fires, or those coming from military attack, were matters of grave concern. Harsh conditions combined with the uncertainty of daily survival left little room for architectural consideration. With survival itself threatened by raids, with disease and social anarchy tearing apart the social fabric, the primary goal of the mason or architect was the completion of the building before the next catastrophe. The integration of architectural detail for existential expression does not keep marauding hordes at bay. This explains why very little masonry architecture of artistic importance was built in Western Europe for nearly five centuries.[16] Architecture is forever whispering the past, and the buildings themselves—or the lack of them—tell us this story.

The advent of hardened metal tools toward the end of the first millennium CE brought about real changes in stone architecture. The combination of different ores to adjust the chemistry and temper of the resulting metal alloy can rightly be seen as one of the catalysts that allowed for the building of Europe's great cathedrals. While metallurgy was primarily fueled by the need for stronger weapons and armor, the advancement of the blacksmith's craft was quickly applied to tooling for other trades. Harder tools enabled a radical shift in stone architecture. Harder hand chisels permitted the processing of harder stone into smaller and thinner pieces. Tightly crafted, geometrically integrated stone could carry more load within a thinner wall

section, allowing the lightening of the stone walls above. Thinner, stronger walls allowed for more penetrations—more light, and more visually open architecture. We see this, of course, in temples and cathedrals, cloisters and inner sanctums, where protection was assumed to be secure. The defensive walls of the various empires remained largely unchanged.

This simple technological shift brought about new worlds of expressive options. It was as if the string quartet that had been making beautiful music for thousands of years had suddenly become an orchestra. New octaves came into range, and new color and shading were added to the experience. The muted expression of the early medieval period, whose solid masses of stone masonry provided shelter and sanctuary, began to yield buildings with the ability to say something more profound.

Of course, culture played an important part in this return to expressive stonework. The end of the early Middle Ages, about 1000 CE, coincided with a new urbanism reflecting increased economic stability. Architecture once more became deeply considered. For the first time since the construction of Constantinople's Hagia Sophia in the sixth century, building forms were consciously attended to, manipulated, and designed.

ABOVE
Some of the earliest uses of stone are so simple, they appear to be part of the landscape itself. In this example, the crafting of the stone stair block is minimal, with just enough shaping to achieve the required function. Over the centuries, the treads have become worn and smooth, bearing witness to their use. Xiamen, China

Architectural forms of the past whisper a narrative of cultural context. This one-room granite church, built in the ninth century, has roofing blocks that corbel unsupported over the central space. Corbeling, unlike arches, may be done without wood centering to support the stones during construction. In many areas, deforestation caused serious shortages of usable lumber, and the little lumber that was available was prioritized for shipbuilding and early metallurgy. This corbeled roof design is successful because the cantilevered stones transfer all the forces vertically, relying on the compressive strength of the granite. This roof survived for more than one thousand years before it collapsed in the early part of the last century. It was rebuilt in the 1970s. St. Macdara's Island, Ireland

In cathedrals and civil architecture, tracery windows, flying buttresses, and soaring vaults supported by tightly knit clusters of precisely shaped stone pieces all owed their invention to advances in metallurgy and the rising unified secularism now joining the ecclesiastical search for human and divine expression.

The scale of this building campaign is hard to imagine. The French historian Jean Gimpel reports that in the two hundred years between 1050 and 1350 CE, "several million tons of stone were quarried in France to build eighty cathedrals, five hundred large churches and tens of thousands of parish churches."[17] As it turns out, more stone was quarried in France during that period than in ancient Egypt during its whole history, and that includes the Great Pyramid, a stone volume exceeding forty million cubic feet. As the buildings were lightened in the beginning of the twelfth century with larger windows and soaring ceilings, the shift to a new style of masonry architecture was well underway. Beyond solving for the practical loads within the building, the stonework once again became a canvas for human expression. Masonry literally reached for the heavens. Gimpel points out that a fourteen-story building would fit inside the choir of Beauvais Cathedral, and the master masons of Chartres would require a thirty-story building to reach their cathedral spire at 345 feet, 6 inches.[18]

This period, known as the later Middle Ages, ran for some four hundred years—from around 1050 CE until the end of the fifteenth century—and marked a dynamic period for Europe. Most scholars credit the explosive population growth and new prosperity to agricultural advances, particularly the invention of the asymmetrical plough with its moldboard to turn the soil. But there were other inventions of equal consequence, including collars for draft animals, the use of dung fertilizer, and switching to a three-year crop rotation. This led to the great expansion of cultivated land, which peaked in the first half of the twelfth century.[19] The time was equally fertile for stonework and masonry's expressive evolution.

By the time medieval culture faded in the fourteenth and fifteenth centuries, individual craftspeople had been actively traveling, exchanging ideas, and working collectively in Europe for hundreds of years. The clearest indication of this type of travel and exchange is found in the remarkable sketchbook of Villard de Honnecourt. Its thirty-three extant pages (it once contained forty-two) are described by G. G. Coulton, a historian of medieval religion: "We find sketches from Cambrai Cathedral and neighboring Vaucelles, from Chartres, Laon, Lausanne, Reims and Meaux. Reims, which was then building, he had studied with especial care, down to the sections of the mouldings."[20] This type of professional exchange proved extremely fertile, and more pure masonry invention was accomplished during this time than in any era up until the beginning of the last century. The advancement in hand tooling, coupled with the relative stability of the culture, brought about this revolution in stone design. If walls could do more than hold up the roof, then the wholesale dressing and shaping of the architectural elements themselves was now the order of the day.

When the new metallurgy—allowing a wider range of expression—combined with an alignment of the social fabric around the shared ideals of medieval Christianity, the results proved dazzling. The combination altered not only the architecture but also the art in and on the buildings themselves. Every surface was now up for reconsideration, and even functional details held expressive potential. If water needed to be projected off a building, why not carve that element into a gargoyle whose mouth was the spout? The previous norm of massive stone walls without large penetrations for windows or doors now begged for surface adornment. With less wall surface for paintings or frescoes, why not sculpt the architectural details directly onto the structural elements?[21] And those larger new window openings? Why not use that space too? The massive stained glass windows of the period were born of this expressive impulse.

Advances in metallurgy give us only part of the story of this moment; the content of what humankind strove to articulate flourished as well. Tooling's technological advancement coincided with a shift in indicative desire. Before the eleventh century, Western Europeans honored God primarily via modesty and self-denial. Saint Benedict's *Rule for Monks* (ca. 530 CE) had set the tone and way of life for countless monasteries over the following centuries. The masonry of the Carolingian period (750–887 CE) echoed Saint Benedict's unpretentiousness and concern for strict functionality with solid massing and understatement, resulting in pleasing but often fortress-like structures. The Benedictine tradition continued all the way through to the Cistercians (founded by Saint Bernard at the turn of the twelfth century). Here, too, we find simple monastic walls without elaboration, barren of sculpture or stained glass. Even the stones were left undressed, their function being merely to keep out the weather and create a gathering space for quiet contemplation.

By contrast, the Cluniac Reforms, enacted a century earlier, led in another direction entirely. Their ideals can be seen easily in the sumptuous life of the famous twelfth-century abbot of Saint-Denis, Abbot Suger, who believed that God should be celebrated with the best that humanity could contrive. The wealth and riches of the world rightfully ought to adorn His home on Earth. Suger's insistence that nothing was too good for God translated into the use of costly materials and the very best workmanship. He had the main altar at Saint-Denis covered in gold, and collected precious stones to ornament the great cross (it measured some twenty-one feet across) hanging in the choir. The cross was covered in "hyacinths, sapphires, rubies, emeralds and topazes" from the treasury of the late King Henry of England, "who had amassed them throughout his life in wonderful vessels."[22] Such newly splendid interior designs and furnishings required significantly improved lighting, and Suger initiated a radical remodel, removing the existing basilica and punctuating the new masonry with windows. Pleased with his efforts, Suger had inscribed in the stone of the church, "Once the new part is joined to the part in front, the church shines with its middle part brightened. For bright is that which is brightly coupled with the bright, and bright is the noble edifice which is pervaded by the new light."[23]

Reflecting the shift to expressive, expensive, well-illuminated furnishings and interiors, building programs themselves changed across the land. No longer was the church a simple vessel enfolding quiet contemplation. It now might serve as a meeting space, an indoor stadium and stage for the increasingly theatrical pageantry developing alongside the modern Christian heritage. These new interior spaces were cavernous. Amiens Cathedral in France, for example, covers 208,000 square feet, allowing nearly ten thousand souls to attend the same service.[24] In Rome, Saint Peter's Basilica, the largest ever built, covers 227,070 square feet (more than five acres) and contains in excess of 163,000 square feet of interior space.[25]

OPPOSITE
Advances in metallurgy and tool hardness altered the expression of stone in far-reaching ways. Smaller, harder stones could be crafted and assembled to create large spaces with significantly more light. This opened up architecture and formed space and light in new ways. Villers-la-Ville, Belgium

Epic Scale

The earliest cultures using the most primitive tools often worked with larger-scale materials. The unknown, pre-Roman builders at the temple site in Baalbek, Lebanon, quarried and shaped some of the largest building stones ever set for the foundation of what would become the Roman Temple of Jupiter. Likely using early iron tooling, this culture was able to quarry and emplace

The foundation stones of the western wall of the Temple Mount in Jerusalem demonstrate a key principle of early stone masonry: strength through mass. The tight joints and extreme volumes create an effectively impenetrable surface that has withstood waves of violence and war since it was first built in about 19 BCE. Jerusalem, Israel three 900-ton blocks to form the raised platform on which the temple was built. In the nearby quarry, there are two additional stones, shaped and ready for transport, weighing 1,200 and 1,500 tons, respectively. These are thought to be the largest building stones ever created.[26]

In early Roman times, with iron tooling and invention of the pulley and compound lifting block and tackle, Herod the Great's craftspeople quarried and set limestone blocks in Jerusalem as large as 570 tons and possibly larger. This same impulse drove the Inca to move massive blocks in the 80-ton range—with at least some weighing upward of 350 tons—up and down the formidable valleys of the Andes.

The great pyramid builders of ancient Egypt routinely squared massive blocks of 30 tons and sometimes far greater. The ashlars (finely wrought rectangular blocks) in the tomb of King Shepseskaf at Saqqara weigh about 220 tons each. Of course, their granite obelisks are among the heaviest stone objects that the ancient Egyptians crafted and moved; the largest weighs 502 metric tons and measures a little over 105 feet in length.[27] This monolith, now known as the Lateran Obelisk—originally set up by Thutmose III and IV in the great temple of Karnak—was stolen from Egypt by Emperor Constantine. As recorded by the Roman writer Ammianus, a special wooden barge of epic scale, rowed by three hundred oarsmen, was built to transport the huge obelisk across the Mediterranean to Rome.[28] Upon arrival in its new home it was erected in the center of the Circus Maximus. It eventually sank into the marshy terrain and wasn't found until the sixteenth century, heavily damaged and burned, suggesting that it "may have been purposefully overturned and mutilated."[29] Today the massive form stands in front of the Archbasilica of Saint John Lateran in Rome.

This copperplate etching details the movement of the Vatican Obelisk in the year 1586. The achievement was celebrated in a fête-book (commemorative event book) by Domenico Fontana, who conceptualized the method and won the pope's commission. The book revolutionized the conveyance of technical information and heavily influenced subsequent architectural publications. Natale Bonifacio da Sebenico is the artist credited with creating the copper plates, working directly from Fontana's own drawings.

A second, smaller obelisk—weighing a mere 365 tons—was also taken and erected at Circus Nero (the present site of the Vatican). According to Pliny the Elder, of all the obelisks transported to Rome, this was the only one to break as it was set in place.[30] Now known as the Vatican Obelisk, when it was moved 843 feet in 1586 CE to Piazza del Popolo after fifty years of planning, it was considered "the biggest material handling operation in one thousand years."[31] Even the reigning genius of the day, Michelangelo, turned down the job on several occasions, steadfastly replying to all entreaties, "Et se si rompesse?" (And what if it breaks?).[32] Ultimately, the move required almost one thousand men, 140 cart horses, and forty-four cranes.

Neolithic cultures of northern Europe erected colossal stone monoliths, called menhir (from the French), most without significant adornment. Such monoliths are found around the world, with more than fifty thousand cited in northern Europe alone. These single stones routinely exceed four tons and often greatly surpass that size. One of the very largest, measuring sixty-five feet in height, rises in Quiberon in Brittany, France, and was erected in the later part of the second millennium BCE. The historian and architecture critic Sigfried Giedion remarks: "The origin of verticality is deeply anchored in mythopoeic thinking. It is the most obvious symbol pointing from earth to heaven—from earthly existence to the abode of the gods."[33] The mythologist Robert Lawlor also hints at a menhir's numinous symbolism: "We can see in the ancient agrarian practice of erecting stone monoliths, the phallic, mineral roots of the earth, the function of attracting downwards the fertile, cosmic ambiance."[34] When there are at least two stones or a chamber, the assembly is called a dolmen. Both phenomena appear to have occurred spontaneously in North Africa, the Middle East, and Southeast Asia, where, as late as the turn of the last century, the erection of menhirs was still practiced by the tribes of the Indonesian archipelago.

Another Stone Age culture, the Rapa Nui (1100–1680 CE), quarried and moved giant volcanic sculptural assemblies called moai across Easter Island.[35] The largest one of these yet discovered weighs 80 tons, and an unfinished version still lying in a quarry has an estimated weight of 270 tons.[36]

Material Scale and Architectural Style

The relationship between material scale and available tooling offers an interesting window into a larger discussion of architectural styles developed by various cultures. Consider that the ancient Greeks consistently used large-scale blocks to create generally small-scale buildings, while the twelfth-century mason-architects constructed enormous buildings with small stones.[37] The Greeks called their style of over-scale masonry Cyclopean, suggesting that only the one-eyed monster Cyclopes could build walls with blocks so large. These walls derived their strength from the mass of the building components.[38] The scale and mass of the Egyptian pyramid blocks were sized and fitted using the same principle: strength through mass to protect the tomb from future assault. In the case of Egypt, the strategy almost worked. Although Khufu's pyramid (the Great Pyramid)

Many of the earliest civilizations worked to erect large stones in the landscape—sometimes for worship, sometimes for pure expression, sometimes for practical purposes. Bedouin tribes have used monoliths as navigational aids for millennia. Erfoud, Morocco was robbed during the Middle Kingdom (2055–1650 BCE) and most of the rest received significant looting before the beginning of the New Kingdom, also referred to as the Egyptian Empire (1600–1100 BCE), it took until the twentieth century for some of the greatest treasures to be found. The tomb of the New Kingdom pharaoh Tutankhamun (ca. 1336–1327 BCE) was not discovered by Howard Carter until 1922 CE.

One might quarry and stand vertical an obelisk of 502 tons—or an Ethiopian stele of 520 tons—to test the strength and resiliency of the team, but also because the ruler's strength must be projected for all to see. For example, the Inca walls of Cuzco have more to do with scale as a projection of power and strength than with any specific requirement of the architectural program. The Inca were using scale to project symbolic power and to intimidate potential enemies. The same principle is at work when a conquering power destroys or appropriates monuments as part of an effort to subjugate and emasculate the vanquished. Nothing projects a rising power better than the capture and removal of another culture's sacred touchstones. Even today, commercial banks use overscaled stone to project power and strength, hoping clients will intuitively believe that their money is safe.

Just as the shift to metal tooling allowed the directionality of impact to be precisely applied, so it propelled innovation in the process of quarrying—removing stone from the earth. In the quarry, tooling is carefully aligned to direct vibration and pressure to break stones free and split them into manageable blocks. When splitting a stone, a craftsperson will typically let the absorbed energy of the hammer blows "set," or percolate, through the material, knowing that the vibrational damage they inflict takes time to reveal itself. The Japanese call this moment of waiting *i pugu*, or *ichi pafu*. This slang term roughly translates as "one puff," meaning that once the line has been scored with the chisel or metal wedges tightened in the block, it is time to roll a cigarette and allow the pressure to build inside the stone; it is never to be forced.[39] It is possible to split apart even a very large stone like a piece of dried kindling, releasing the energy created by the pressure and vibration of the tooling as it works through the material over time.

But this vibrational damage occurs in other, unexpected ways as well. In what I call the "torture of the craft," unintentional damage or internal cracking of the material may not be obvious during the first phase of fabrication, but may show up later, during the added shocks from shaping blows. Every dimensional

stone carver and sculptor has a story of a work, almost roughed out, suddenly cracking apart without obvious cause. Such experiences can be psychically shattering as well. Still, once a stone is installed safely with carving and manipulation complete, the finished work virtually never changes—cracks or breaks—without shifts from external mechanical forces.

Knowing that hidden fractures can reveal themselves during fabrication would seem to argue for keeping the material as large as possible for transfer to the building site. Yet we know this was not done. Almost all cultures performed the primary shaping and fabrication at the quarry location so as to reduce weight and therefore the effort and cost of transfer or freight.[40] Stone fabrication is, after all, a reductive process. Every manipulation of a large block makes it smaller, so why pay to move a large block if some significant portion of the material is bound to be lost in the fabrication process?[41] The solution is to pre-craft the material at the quarry as close to finished size as is practical. That said, these considerations must be weighed against the reality that although thinner pieces may be easier to move and manipulate, they are also more likely to break in transit. Once pre-sized pieces arrive at a site where the vibrational risks are much lower, further shaping and fine-tuning invariably takes place.

ABOVE
Contrary to expectation, extremely large slabs may be created using simple hand tools with vibration and directed pressure. Beaverdell, British Columbia

Of course, not all quarries produce large blocks. Many times, the stone shell of the Earth is fractured into narrow veins of weakness. Most of these fissures are cooling and contraction cracks from the geological events following the stone's creation. Although the Earth is estimated to be 4.8 billion years old, more than two-thirds of its surface was "thrown up" in the last 200 million years through volcanic action. Most of the rest of the cracks that limit a quarry can be traced back to the estimated 150,000 annual earthquakes that the Earth's surface endures.[42]

In this way, the size of the raw stone block is directly linked to the breadth and depth of the un-fractured, living rock available to be quarried. If great strength is required, a quarry capable of providing blocks of great scale must be located and developed. Or, putting it another way, since the scale of a piece is directly related to the ultimate strength of the object created, the scale of the available material resource becomes a key driver in material selection. The Egyptian obelisks are pink in color—not khaki to match the walls of the Temple of Luxor—because only the quarry on the Nile at Aswan could provide such lengths without internal fractures. In fact, the Aswan quarry remains one of the outstanding formations of un-fractured living rock on the planet ("living rock" is a quarry term for undisturbed rock with the groundwater still flowing through it; more on this soon). Had the ancient kings of Egypt not fancied the pink color of the Aswan stone, they would have been forced to build much smaller monuments to their power and prowess.

The Science of Stone

The scientific knowledge of stone is astonishingly recent. If the roughly five-thousand-year history of stone architecture were compressed into a two-hour movie, our scientific knowledge of stone would occur in the last sixty seconds. Most of what is known about what stone can and cannot do has been empirical knowledge passed down from the temples of Egypt, through the medieval guilds, to us today. We can see how the earliest stone cultures struggled to work this knowledge out when we contemplate the first columns at the tomb complex of the Step Pyramid at Saqqara. It is ascribed to Imhotep, the builder-architect of Pharaoh Zoser (sometimes spelled Djoser or Djeser), credited as the inventor of the art of building in hewn stone.[43] In the middle of the west side of the entrance hall we find the world's first stone columns articulated as distinct architectural elements. They are even decorated with vertical flutings that accentuate the grandeur of the form (much the way vertical stripes on a sweater can make us look taller). But the column remains engaged to the wall, not yet independent, as the bearing capacity of the new form was not yet understood.[44] The column continued to be empirically refined in Greek architecture, where the column, now separated from the wall and freestanding, became slenderer with time's passage.[45] As the architectural historian William Bell Dinsmoor noted, the Greeks maintained "the traditional Greek suspicion of the strength of stone" and worked to strengthen it with iron clamps and bars.[46]

The first stone columns remained connected to the wall behind, as builders cautiously explored their weight-bearing capabilities. Note the vertical flutings that visually exaggerate the columns' height. Saqqara, Egypt

Similarly, consider the common sloped doorways that appear in virtually every ancient culture. Academic critics will argue that this slope is meant to counteract the visual bending of vertical lines in the same way that entasis was employed to counteract the visual swelling of vertical columns.[47] However, to a stone practitioner, the more obvious interpretation is that the builders were simply trying to reduce pressure on the spanning lintel. By narrowing the top of the opening, the distance of the span is diminished, reducing the chance of failure. Lacking modern testing or other technical means of verifying the strength of a spanning member, the builders intuitively understood that their chances of success were considerably raised by both increasing the thickness of the lintel and reducing the distance spanned.

Inherently cautious by nature, early builders imitated the materials over which they had developed more confident mastery. We can see early stone columns expressed as bundled reeds—a successful copy of an early solution in which a column is created from organic matter.[48] This perceptual strength would add nothing to the actual strength of the assembly, but it would certainly read as stronger to an untrusting citizenry or ruler. Likewise, early stone vaults carved into living rock often imitated traditional wood forms. A ceiling or vault carved with completely irrelevant coffered beams may have provided the illusion of strength to those praying and taking sanctuary within. Imhotep put a flat stone roof on the tomb at Saqqara, complete with beams "carved on the underside to imitate rounded logs of wood."[49]

In some cultures, and with some materials, limitations related to size and strength were sidestepped by carving an interior space directly into living rock. This technique works beautifully in soft materials such as the tuffs (a consolidated volcanic ash) found in Italy or compact sandstones common in many parts of the world, including Jordan, Turkey, Ethiopia, Afghanistan, and China. The

The inclined doorway, invented independently in many ancient cultures, reduces the spanning distance required of the lintel at the top. Sacsayhuamán, Peru

TOP
Ollantaytambo, Peru

BOTTOM
This spanning lintel is supported by projecting blocks—an inventive solution that reduces the weight and strain on the lintel to decrease the distance spanned without tapering the side walls. Cape Town, South Africa

OPPOSITE
At the Temple of Luxor, columns suggest bound reeds, a proven local building strategy that predated stone's use. In this way, early stone builders appropriated the perceived strength of trusted materials already widely utilized. Luxor, Egypt

ABOVE
This fifth-century temple was carved directly into the living rock. Elaborate "wood" beam forms visually support the vaulted ceiling. Of course, this is an illusion, since the entire structure—columns, altar, and coffered ceiling—is carved from one contiguous lithic mass. Ajanta, India

The Leshan Buddha was carved directly into the living rock in the eighth century by Buddhist monks. Leshan, China

Romans carved the hard sandstone at Petra with elaborate classical detail inherited from Greece. Working directly on the living rock with iron tools, they achieved crisp arrises and precise geometries. Inside the structure, the cavernous room is kept cool by the base temperature of the mountain, offering respite from the searing desert heat. Petra, Jordan technique is even found in harder limestone formations along the Nile in Egypt and in some of the underground pyramid vaults.[50] But most surprising of all are the hard basalt formations of western India, where basalt, one of the most difficult and intractable stones on the planet, was worked to perfection with early iron tools.

Carving directly into living rock is still done in modern times. The epic sculpture at Mount Rushmore in South Dakota and the ongoing creation of the nearby Crazy Horse Memorial are excellent examples. In both, the coarse-grained granite is shaped with careful drilling, in combination with explosive blasting, to remove material and bring the living rock into sculptural form.

Transportation

The problem of transport has been a thorny one that every civilization building with stone has struggled to solve. Overcoming the physics of moving heavy things is complex and is never inexpensive, and thus transportation has an outsize impact on the schedule and budget of any stone project. As early as the Sixth Dynasty in Egypt (the last reign of the Old Kingdom, 2345–2181 BCE), a governor known as Una described his method of conveyance of stones for a pyramid: "I made first a boat of burthen, 60 cubits long and 30 cubits broad [110 by 55 feet], as soon as the water rose, I loaded the rafts with immense pieces of granite for the pyramid."[51] Almost 350 years later, King Mentuhotep IV of the Eleventh Dynasty sent ten thousand men to retrieve a single stone for a large sarcophagus.[52] In another example, it took a mere three thousand "sailors from the Delta Provinces" to remove the lid of a different sarcophagus, a stone measuring approximately thirteen by six by three feet.[53] Approximately 750 years later, in the Nineteenth Dynasty, Ramesses II floated an obelisk down the Nile during the annual flood on barges of river reeds "by the exertions of 120,000 men." He reportedly "bound his own son to the summit of the shaft, during its erection, so that the officials might not neglect its care."[54] I've certainly faced some job stress in my career, but this example is in a class by itself.

Shipping stone by water has proved to be a timeless strategy. The historical record is replete with interesting citations; for instance, the British writer Tom Williamson mentions that the "great *marani*—ships that could each carry a load of up to two hundred tons—brought delicate, white, marble-like stones from quarries on the Istrian peninsula in what is now Slovenia, but at

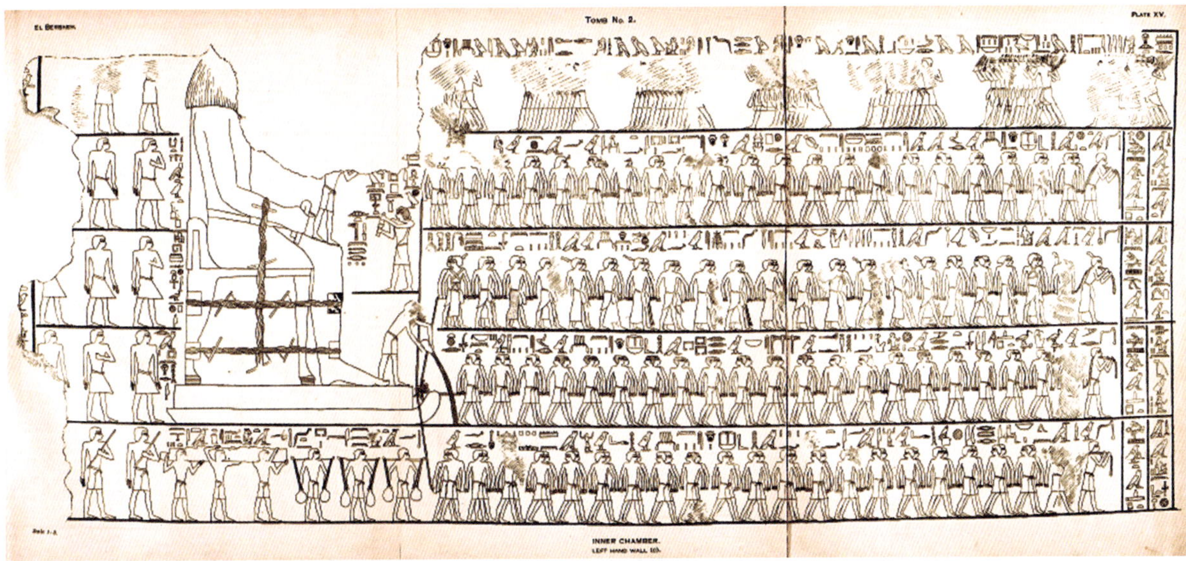

the time belonged to La Serenissima [Most Serene Republic of Venice]."[55] Worked stone was also used as ballast to pack around goods after sailing ships were loaded. Cobbles and paving slabs were particularly prized, since their relatively small and dense nature allowed them to be fitted into odd cavities of tall ships to achieve an ideal draw and waterline. New York's Wall Street is paved with golden- hued granite planks from Amoy, the old name for the city of Xiamen, China, in Fujian province. Black limestone pavers from current-day Belgium can be found in Indonesia, remnants of the Dutch colonial shipping trade. Another relatively modern example of stone hitching a ride to faraway places is the black basalt of San Francisco's street cobbles, originally from the "bluestone" of Melbourne, Australia.

Shifting heavy stone for even short distances can be problematic. The obelisks were loaded using clay mortar as a friction-reducing agent to slide the massive stones into position. The Egyptian tomb of Tehuti-Hetep (or Djehutyhotep), a Twelfth Dynasty monarch, contains a wall painting that depicts the transportation of a colossal stone statue on a sledge. It is being dragged by several hundred men, while a few in front pour a liquid over the roadbed. It has been suggested that this is tallow (heated animal fat) or olive oil "to harden the sand and facilitate the dragging."[56] The building blocks for the Anji Bridge, China's oldest standing bridge, were moved in winter when they could be slid over ice. Peruvians used cactus slurry as a lubricant on which to slide heavy blocks.[57]

ABOVE
Friction-reducing agents were often used in the transport of large stones. This reproduction of an Egyptian tomb wall painting shows craftsmen pouring liquid ahead of the skids on which the stone colossus is mounted. Most scholars believe the liquid is oil. Other cultures have relied on a variety of lubricants, including clay, cactus slurry, and even ice.

Clearly, none of this shifting and transport of stone was free. The building historian and author Malcolm Hislop reports, "At Caernarvon Castle the total amount spent on materials in 1285–6 was £151 5s 6 1/2d but the cost of transportation was £535 8s 8 1/2d [3.5 times the cost of the material]."[58] Of course, during the Industrial Revolution, expanding barge canal routes and the use of steam locomotives eased the transport of materials to growing cities, significantly lowering costs. Similar changes influenced fabrication. As the water mill gave way to steam power, the processing of stone was finally uncoupled from the source of the processing power. Unrestricted by proximity to rivers and streams, crafters of stone relocated to either the quarry (the location of the material) or the city (the location of the clients). These advances significantly fueled the growth of the stone industry. Cable cranes, stone lathes, gang sawing, and, finally, pneumatic drills used in quarrying reduced cost by speeding extraction and production.[59]

Curiously, these newfound technologies seem also to have fueled a nostalgic impulse to revisit past architectural vernaculars. More often than not, better tooling meant freedom to make more classically styled elements, including column capitals and balustrades. A more troubling development was that better tooling meant many spectacular objects of past architectural glory could be made thinner and lighter, their original proportionality altered. Proportionality is fundamentally related to the visual credibility of an object. From a practical perspective, the proportion of a granite column—given the superior strength of the material—may naturally be different from one created out of sandstone. Historically, the crafting of a sandstone column with hand tools restricted the possibility of making the column too small (and therefore out of proportion), since the percussive action of the hammer and chisel would simply have broken the material apart. In the late nineteenth century, however, the cutting action of new tooling made carved forms available beyond what the traditional strength of the material might have otherwise allowed.

Another example can be found by studying the stone balustrade, since advanced tooling made it possible to cut the neck of the form down to as little as two inches in diameter, a narrowness previously impossible. Yet this processing improvement does not make such a balustrade more visually credible. Rather, the opposite occurs: the balustrade's narrower neck is both structurally and visually weak. Consequently, modern tooling allowed this form to be altered in such a way as to corrupt our belief in the material itself. We might point to this as the moment when the safety and strength historically associated with stonework began to be distorted by injudicious use of advancing technology. As we will see, these new capacities had far-reaching consequences.

Paul Wiseman, one of the twentieth century's great interior designers, once remarked to me, "Just because you can, doesn't mean you should," illuminating the strong relationship between design and restraint.[60] Most of the tooling breakthroughs of the past hundred and fifty years have thrown such restraint in stone crafting out the window. Perhaps predictably, modern society's first instinct was to mechanize stone fabrication in order to lower unit costs, sell more stone, and make more money. Fabricators reduced the labor involved in creating stone objects from the past rather than exploring how these breakthroughs might expand stone's future visual range.

To be clear, reprocessing the past is not in itself a bad thing. At its best, it creates a kind of socialism of architectural detail, since those previously unable to afford the labor of complex moldings and fine stone articulation suddenly find such luxury within reach. The sad consequence, however, is that producers of these details, often at the request of the designers, often pull and stretch the proportions of the objects into abject meaninglessness, modifying the historical "product" to suit the new tooling or packaging rather than vice versa. Many stone products now come in a maximum length of thirty-two inches, since that size best fits on a pallet. Perhaps, on this subject, we leave the final word to the Dominican friar and philosopher Thomas Aquinas, who reminds us: "The senses delight in things duly proportional as in something akin to them; for, the sense, too, is a kind of reason as is every cognitive power."[61] Looking to delight our senses may yet hold the key.

Modern Methods

Fast-forwarding the aforementioned hundred and fifty years to the present, we find stone's fundamental expression once again altered by technological advances. While changes resulting from the Industrial Revolution brought many iterative improvements to stone processing and transportation, it was not until the introduction of low-cost industrial diamonds coupled with space-age epoxy resins that the first expressive shift of significance since the twelfth century found traction. Combined, these new technologies have allowed building stone to be deployed in highly improbable locations. For example, stone laminated to aluminum or resin honeycomb backing is now used extensively on personal and corporate jets and yachts.[62] Modern mortars combined with powdered epoxies are now capable of adhering stone to glass. Modern engineering coupled with significant developments in caulking and flexible jointing systems have pushed stone into the heavens. In Chicago, stone cladding can be found on the Aon Center, which was for decades the world's tallest stone building, at 83 floors and 1,136 feet.[63] Most recently, even that record has been surpassed by the Al Hamra Tower (1,354 feet, but only 80 floors), Kuwait's tallest building. In that case, by leveraging advances in epoxy technology, the limestone for the upper floors was adhered directly (mesh-mounted) to the swaying substrate, essentially installed as tile.

By many measures, the practice of stonemasonry has evolved more in the last hundred years than in the previous five thousand. Alongside this, logistical advances in modern container freight have brought unprecedented access to stone materials quarried around the globe. To put this in perspective, at the end of the nineteenth century, there were approximately 350 kinds of stone available on the world market. By the year 2000, there were more than 5,000. Greater availability has naturally fueled a wider range of applications. Data for 2018 revealed that total US consumption of stone was just over 3 million tons, having more than doubled in twenty years.[64]

But this is only half the story. The real difference lies in how stone was used prior to World War II versus how it is used today. For nearly the entire history of building with stone, the average thickness of material used to construct a building ranged from eleven to thirty inches. Not only were the pieces sized larger, but the walls assembled of these stones were themselves massively thick. Hardly atypical, for example, are the covered corridor and causeway walls at the Pyramid of Unas (Fifth Dynasty, Saqqara), estimated to be eighty inches in thickness.[65] Closer to home, the masonry walls of Philadelphia's City Hall, built between 1892 and 1901, measure an astonishing twenty-two *feet* in thickness.[66] Yet during the last half of the twentieth century, average stone thickness of walls dropped to less than two inches, with supporting walls almost always made of another, non-masonry material. As building technology has continued to evolve, wall thickness has continued to plummet, and most of the dimensional stone deployed thus far in the two decades of this century measures as little as one and one-quarter inch in thickness.[67]

Each technological advance allows stone to be used in new ways. Economical ways of cutting it into slabs of consistent thickness was one type of revolution. However, it was not until this technology was combined with other breakthroughs that the magnitude of the changed expression becomes apparent. Attaching thin stone slabs to a steel substructure via flexible, waterproof connections fundamentally altered stone application for the first time in human history. Suddenly, stone is not holding up a building but rather being used as low-maintenance rain screen—a wallpaper, really—to separate the inner core from the outer skin.

At the same time, the enormous efficiencies of global container freight have effectively broken the competitive lock that the material closest to the project enjoyed for centuries. In terms of building stone resources, the promise of a "global village" has become a reality: stone may be quarried in Brazil, China, or South Africa, shipped to Italy for sawing, shipped to Norway for attachment, and finally shipped to New York City for installation. It often costs the same to ship a container of material from Hong Kong to Los Angeles as it is does to drive that same container by truck from Los Angeles to Seattle.

This revolution in transportation has been exciting. Every modern city's stone architecture now reflects the diversity of its world citizenry with a breathtaking collection of world geology. How different this is from the past, when the white marble of Athens, the travertine and pozzolan concrete of Rome, and the beige limestone of Paris were so essential to each respective city's identity. Urban centers coming of age after the invention of modern transportation have effectively been released from their regional geology. Although the change irrevocably altered the enforced continuity of the past, it has also expanded the palette available and added undeniable vibrancy to our urban architecture.

The sudden changes in stone processing technology and structural requirements have left many highly competent design professionals bewildered, and no wonder. The long-established masonry tradition and knowledge transfer in Europe were in fine shape until World War II. But the epic suffering and destruction of that conflict, with its estimated sixty million dead, produced a profound disruption to the Western building tradition that is hard to overstate. After the war, the urge to build a more utopian future drove modern architecture away from traditional building materials. The future was to be constructed of "modern" materials, factory controlled and precise, free of natural variation and subtle nuance. Steel and glass, machined to perfection, became the fashion. Handcrafted, expressive materials became associated with a broken and damaged past, a marred legacy of the prewar era. Most of the sophisticated stonework performed in the aftermath of the war was in the rebuilding of what the war had damaged rather than the creation of new stone structures.

This impulse to break with the past in order to reinvent the future produced some devastating results. One of the greatest stone buildings in the United States—Pennsylvania Station in New York—was torn down to make way for the architectural nightmare of Madison Square Garden. Fine monuments of the past were "remodeled" with T-bar drop ceilings, obscuring domes and/or masonry details.[68] In China, the same impulse was given explicit state sanction under the ominous title Great Leap Forward. As the peasants melted down their plowshares at the urging of the central government, an estimated thirty million people died of famine in Sichuan province alone. The stone sculptural objects of Chinese history were systematically destroyed to focus the public imagination on the future. The inclination to break with the past helps explain why postwar architecture contains so little stone that isn't machine finished, and thus largely neutered of its potential expression.

With war and fashion conspiring to dislodge stone from the pinnacle of the built environment, the body blow from engineering could not have come at a worse time. Stone, with its infinite natural variation, ultimately came to be classified as "nonstructural" by the emerging science of engineering. This fate was shared by a number of other materials whose extensive use over the millennia provided what we might fairly imagine as the gold standard of structural success. Even heavy timber was ruled "statistically indeterminate" and, in most jurisdictions, no longer allowed to be used structurally. These radical conclusions, particularly concerning stone, have been hard to comprehend for many traditional masons, since up to this time the world's largest structures had been assembled of masonry, and the structures often survived the assaults of both time and humanity. Indeed, anyone who has marveled at the beauty of a flying buttress on a Gothic cathedral may feel impassioned to argue for stone as a structural element. Or consider our continued use of the Pons

Fabricius, the oldest existing bridge in Rome, which so elegantly continues to span a section of the River Tiber after more than two millennia of daily use. This Roman bridge contains no structural steel, instead relying solely on the compressive strength of the wedge-shaped voussoirs. Yet central Italy has been consistently wracked by earthquakes that cause considerable loss of life and widespread devastation.[69] The friction-fit voussoirs provide a timeless solution, but only as long as the earth on which they rest remains still. Although the stiff banks supporting the Pons Fabricius have remained stable through the rise and fall of civilizations, when—not if—those anchoring banks waver, the bridge will finally fail. Clearly, neither the engineer's nor the stonemason's viewpoint is explicitly wrong, but rather raises a larger question: To what timescale are we building?

Faced with what may well be an unanswerable question, it is useful to remember that engineering remains an art as well as a science, and a young science at that. Engineering, as practiced by a discrete group of professionals, is a relatively new phenomenon. In the same way that designers were builders until the Renaissance, architects were engineers until the 1920s, when that specialization was finally articulated as its own arena of professional expertise. It was perhaps inevitable that this new science, grounded in hard math and groundbreaking material analysis, should come to be based on a flat-out rejection of anything not predictably quantifiable. Shocking as it sounds, engineers have concluded that stone, the dominant building material of the last five thousand years, cannot be relied upon structurally, based on the inconvenient fact that it is not entirely uniform. Stone's natural inclusions as well as hidden contraction fractures make it vulnerable. Even stone's figured veining is considered a weakness under tension and shear. Short of testing every individual piece to identify its hidden faults, stone's essential variability makes its structural value inconsistent, and buildings constructed from it vulnerable.

The window on the use of structural masonry continues to close. The art historian Erwin Panofsky once remarked that the endings of historical phases "do not as yet add up to a fundamental change in attitude but rather manifest themselves in a gradual decomposition of the existing system."[70] While the transition to fully nonstructural masonry is incomplete, the die has been cast, the ascension of the engineering profession in our age of ultra-rationalism perhaps inevitable. As in the past when we called them masons or architects, modern engineers produce

Engineering advances in the last hundred years have brought the structural limitations of stone into sharp relief. The collapse of inhabited masonry structures, even those of just a single story, almost always prove fatal to humans inside. Clansayes, France work that continues to dazzle us as it lightens and extends structures beyond previously known limits.

With this in mind, let us remind ourselves that the emerging science of engineering, as applied to stone masonry, has already saved tens of thousands of lives. The elemental qualities of "stone-ness"—weight and mass—are inarguably attributes that crush and maim human life when the material collapses or comes unexpectedly undone. The state of Gujarat in India experienced a significant earthquake (magnitude 7.7) in January 2001, in which an estimated 16,500 people perished in the stone rubble of structures that rarely reached even three stories.[71] Even for the survivors, it was a devastating experience. I was nearby, having left the epicenter at Ahmedabad the night before for a last-minute early morning appointment. Even three hundred miles away, my six-story masonry hotel swayed, cracked, and groaned under the strain. It was terrifying. Contrast that result with the San Francisco Bay Area earthquake of 1989 (magnitude 6.9) that killed 67 people. Although the disparate magnitudes of these quakes make direct comparison difficult, the SF quake involved variables that might have made the loss of life more dire: multi-story masonry buildings, large tracts of development on landfill, and bridges over huge expanses of water. Yet modern building codes prevented the catastrophic loss of life that was experienced in Gujarat. When the earth moves, even a one-story home, if made of unreinforced masonry, will likely devastate its inhabitants.

That stone is now a nonstructural entity has yet to be fully absorbed by the building industry or by professional designers. Working and providing stone materials and design around the United States in various seismic zones and climates reveals enormous variety in local practice despite the convergence of building codes on this issue. Even when codes classify stone as nonstructural, stone is often being applied structurally in local practice. At some of the top architecture and design firms, I still get basic questions on this point. Many architects assume that the reason we have turned to nonstructural applications of stone is cost, confessing that they "long for the day when a client has the budget to build a full-depth masonry structure." Such a statement illustrates that many of our most qualified professionals have yet to grasp stone's reclassification. Paradigm shifts are infamously slow to be absorbed. Defining stone as nonstructural means that it is not allowed to do anything more than hold itself up. It may protect the building from wind, water, and sun, but it cannot assist in holding up the windows, roof, or interior walls. In a radical alteration of the rich tradition extending nearly fifty centuries into the past, this is a strange and uncomfortable state of affairs.

Nonstructural stone is technically called veneer or cladding. Veneers are as old as stonework itself and are not, in and of themselves, a bad thing. The original veneer application comes to us from ancient Egyptians, who wrapped the structural cores of their pyramids in finely wrought granite twenty inches thick. In another famous example, Herodotus mentions the causeway at the Great Pyramid, covered in bas-relief carving. The Egyptologist I. E. S. Edwards describes it as composed of local stone "faced on the inside with red granite.... Outside, the lowest course was also faced with granite, but the remainder of the building had a facing of Tura limestone."[72] The Romans used veneer extensively, wrapping their pozzolanic concrete structures in fine marbles inside and out. Eugène Emmanuel Viollet-le-Duc complains bitterly that their common brick or concrete buildings received "a decoration of marble which has no absolutely necessary connection with that edifice,"[73] whereas in Greek architecture, asserts the US architectural scholar Edward R. Ford, "decoration and structure were inseparable."[74] Medieval masons also commonly assembled buildings of simple blocks (brick or rough stone) wrapped with veneers of finely crafted stone. It is the combination of stone veneer with non-masonry structural materials, such as steel or wood, that has fundamentally altered the equation in modern times. As early as 1929, the British educator and early engineer E. G. Warland offered a progressive summary of the issue: "The introduction of steel into the construction of buildings has revolutionized the orthodox methods, and we find that masonry, instead of being the art of building in stone, has become the means whereby the carcass of the building may be pleasingly covered with stone."[75]

Given the constraint that stone is now only allowed to support itself, we come to recognize the complexities of attaching it to the building skeleton or actual structure. Complicating factors include the divergent rates of expansion and contraction between diverse materials—between a heated interior and a potentially frozen exterior, or an air-conditioned interior and a blistering-hot facade. How are we to integrate the "system" of the exterior stone veneer with a structure and waterproofed core designed to expand, contract, and sway? How to create attachment solutions that, while safe, are also cost-effective?

These technical questions exceed the scope of this book, yet they are central to the continued utilization of the world's most beautiful building material.

Although the conclusions of modern engineering and science are painful, I do not dispute them. Rather, I choose to focus on the challenge that lies before us: how these decisions may lead to a renewal of our relationship to the magnificent material that is stone. To do that, we must first acknowledge that while our affection for stone may remain, the many millennia in which it provided our shelter, and kept us safe and dry, have now formally passed. No one wants to abandon stone. Our use of it remains elemental to our sense of safety, solidity, and connection to Mother Earth. In many developed cultures it still communicates financial success and permanence. It remains one of the primary visual landmarks to communicate our legacy and contribution to the world after our deaths. The challenge, then, is to embrace the new reality of nonstructural stone while finding ways to express our design goals within its limits. We do this by returning to stone's essential qualities, and by returning to a time when the material dictated the use, and the use dictated stone.

Chapter Two

GUILD KNOWLEDGE

The rules of masonry haven't changed because the rules of gravity haven't.
—DONALD FRIEDMAN, PE[1]

It was through my apprenticeship in Siena that I began the real work of ascertaining the iterative experiments that were the guild's source of mastery. As I would learn, the guild was heir to an unbroken chain of successful stone strategies extending all the way back to the earliest architectural forms arising on the alluvial plain of the Nile Delta. This remarkably simple rule set is formally known as the Sacred Rules of Bondwork (or *regole sacre* in Italian) and was indeed considered sacred. Indeed, the rule set's value was considered so great, it constituted one of the most closely guarded compendiums of knowledge ever held by a human society. The Rules of Bondwork are as follows:

Lay all stones in their natural bed (never on their points).
Weave it together for strength.
Fortify ends and penetrations.
Build on a good foundation:
    Foundations should be one-third as deep as the structure is to be high.
    Stone supports wood.
Scale it appropriately to need.
Select each stone to its best use.
Make it sweet to the eye.

In quantity, the amount of knowledge the Rules contained was small, and thus controlled with relative ease, but as a basis of political and economic power, its material importance was frequently large and decisive. Consider: if my walls and fortifications are stronger than those of the next tribe or kingdom, then my survival is more assured. If my bridges are more robust and able to span greater distances, then my ability to trade with my neighbors is enhanced, and my economy more likely to flourish. We know for certain that Roman engineering, with its construction techniques undergirding a vast network of roads and bridges, contributed mightily to the success of that vast civilization. In a purely Darwinian political model, every infrastructural success increases the likelihood of survival and expansion. The formulas and techniques underlying such success were the state secrets of their time. Again, we know for certain that in ancient Egypt, more than a millennium before the Romans, the earliest masonry secrets were gathered and guarded. The Danish engineer and geometer Tons Brunés described them as "an occult geometric system which existed as a hidden and sacred factor" closely held by sanctified groups. Such secret societies "could almost be regarded as a type of trade union bound to protect the trade from outside imposters by keeping secret the various procedures used in their work."[2]

Origin of the Rules of Bondwork

We can glimpse the transfer of this sacred knowledge in extant scraps of ancient recipes and formulas, beginning with a Greco-Egyptian papyrus found in a tomb at Thebes dating from the third or early fourth century BCE. Other shards of material knowledge have been preserved in a manuscript in the cathedral library at Lucca, Italy, copied in the eighth century from Greek originals. Astonishingly, the Italian manuscript contains several of the same mortar and plaster recipes that survive in the Greco-Egyptian papyrus.[3] These formulas, along with many others, appear again in the tenth-century treatise *De Artibus Romanorum* by Eraclius, as well as *Schedula Diversarum Artium* (A Little Scroll of Diverse Arts) by the twelfth-century German monk Theophilus.[4] And, most tellingly, German guild meeting notes from a 1459 conclave in Regensburg offer hints of the hidden knowledge that ultimately would become the Ordinances and Articles of the Fraternity of Stonemasons, known as the *Brother Book of 1563*.[5]

Even while the secrets themselves might be glimpsed from time to time in written records, more commonly the surviving references are about the responsibility of keeping them well hidden. The *Brother Book of 1563*, for example, states:

> Every apprentice when he has served his time, and is to be declared free, shall promise to the craft, on his truth and honor, in lieu of oath, under pain of losing his right to practice masonry, that he will disclose or communicate the mason's greeting and grip to no one, except to whom he may justly communicate it; and also that he will write nothing thereof.[6]

The "secrets of the temples" were primarily communicated orally, rather than committed to a script that an outsider could discover. Indeed, "if a temple brother received any teaching from the inner circle he had to pledge solemnly never to reveal the secrets to any outside party."[7] Throughout ancient and medieval times, the knowledge of stonework was thus both kept secret and

Depiction of a masons' lodge in a medieval stained-glass window. Lodges were considered "places of work and rest and of secret conclave." [Francis B. Andrews, The Mediaeval Builder and His Methods (1925; repr., New York: Barnes and Noble, 1993), 10.] Chartres Cathedral, France

transmitted, finally ending up as the main responsibility of the Freemasons guild, which had to carry it forward safely. As the guild historian Francis Andrews explains, "Both the master and the workmen were members in and under the rules of a guild, and they were entirely controlled by conditions determined by the craft, irrespective of any particular place or work; that they were sworn to carry out work according to those rules, even down to the moulds and patterns they used....And further that in connexion with these men there were lodges or places of work and rest and of secret conclave."[8]

At least during the medieval period, membership in this elite profession was open only to the sons of "freemen," those who had gained their freedom from the claims of land or town through birth, good deeds, or a profession.[9] They were assigned to a journeyman mason, and over many years of apprenticeship they eventually became journeymen themselves.[10] If their skill proved sufficient, in time they might earn the classification of master mason.[11] The qualifications for this rarefied title were considerable. The journeyman would have been required to show mastery of tools, techniques, mortar, and chemical formulas, "even down to the moulds" (the geometric profiles used to create moldings) and patternmaking (the manifestation of building geometry).[12] Journeymen would also have had a baseline understanding of plane geometry and the creation of core geometric elements, including the isosceles (right angle) and equilateral triangles and the twelve orthogons, the pathway to the first "secret" of consequence: the ratio known as the golden section. These skills mastered, a journeyman or "fellow"[13] would then travel abroad for a period of one to three years, working under masters in other countries. Finally, the last stage of training for a journeyman required two years' work in their home country from a chosen master.[14] These master masons would impart the final tenets of the craft: the ability to draw an elevation from a building plan,[15] detailed knowledge of the systems of proportion,[16] and critical practical experience with the entire canon of the Sacred Rules of Bondwork and the Sacred Geometries. In addition, master masons were required to have intimate knowledge of the ratios and formulas used to calculate the stresses of buildings. These calculations, though unsophisticated compared to modern mathematical models, were founded on long experience and careful observation, and they were sufficient to enable the construction and continued existence of such masterpieces as the Duomo in Milan and the Hagia Sophia in what is now Istanbul.

Master masons created masterworks. Trained as builder-architects, they were both master builders and master designers of structures. Often they were also entrepreneurs and artists, serving those at the top of the social and ecclesiastical ladders.[17] As the best paid and best organized group of workers in Europe,[18] they took great pride in their accomplishments and often signed their work. They could travel freely, make binding legal contracts, and fine, punish, and maintain order within their organizations.[19]

Guarding this *arcanum magistri*—the esoteric building science of the masters—was a sworn duty, and there were significant consequences for those who betrayed the secrets. In 1099, the Bishop of Utrecht was murdered by a master mason when it was revealed that he had coerced the mason's son to disclose the method and geometry of laying out the church foundations.[20] R. A. Schwaller de Lubicz makes clear the importance of such mysteries in his review of ancient temple secrets: "Did not Pythagoras wait twenty years before being admitted into the Temple? Did he not, in his own teaching, impose silence on pain of death? Therefore, this teaching was not written down.... Herodotus often mentions the obligation he was under to remain silent concerning 'sacred' subjects. Therefore, these instructions had not been changed."[21]

We do find small bits and pieces of the masons' secrets published in the late fifteenth and early sixteenth centuries. The most famous is in the sketchbook of Villard de Honnecourt mentioned in chapter 1. Further fragments were published by German master masons dating from the same period.[22] But more typically, the rule set was not shared publicly but handed down in secrecy through the guilds themselves.

If one danger for secrets is that they might be revealed, in this case a sad but greater danger is that they might be forgotten and lost. As it turns out, humanity's technical and architectural march has been far from linear; basic geometric forms have in fact been invented and then forgotten. The fragility of human structural knowledge and technique can be illustrated by the emergence of civilization in southern Mesopotamia. There, archaeologists have found architecture dating to the mid-third millennium BCE that used both the true arch and the Gothic, or pointed, arch incorporated into vaults and domes roughly 3,300 years before the Romans took these forms mass market.[23] The pointed arch appears again, much later, in a vastly more visible location at the Dome of the Rock in Jerusalem. Originally built in 670 CE and still standing today, its structural techniques were employed at

least five hundred years before the Gothic arch became an icon of medieval architecture in Western Europe.

Recipes and essential formulas can be lost, but they are also sometimes rediscovered. For example, the Roman recipe for concrete, lost during the fall of the Roman Empire, was not reinvented or rediscovered until 1824 in southern England, almost 1,100 years later.[24] Even building materials and technologies can disappear forever. One of my favorite examples is the flexible and ductile glass reportedly invented in Rome at the time of Tiberius Caesar (42 BCE–37 CE). When it was presented to the imperial ruler, he threw it to the ground in an attempt to break it, and the thrown glass reportedly bent instead of shattering. Caesar is said then to have inquired who else might know of the secret material. When it was sworn that no one else was so privileged, he ordered that the inventor be killed, lest the value of gold and silver turn worthless in the face of such advancement.[25] This story was, as the historian Pliny wrote, "more widely spread than well authenticated."[26] The recipe for ductile glass, if it ever truly existed, has never been rediscovered. Still, with today's wider understanding of resins and plastics, the reported invention appears more credible.

ABOVE
The start-and-stop nature of human knowledge is illustrated in southern Mesopotamia, where archaeologists have found architecture dating to the mid-third millennium BCE that uses both the true arch and the pointed "Gothic" arch and incorporates vaults and domes. Luxor, Egypt

The recipe for ordinary glass was itself nearly lost. The ruins of Pompeii preserve panes of Roman glass, and a pane of crown glass (made using a method of spinning glass into a large sheet) dating from the fifth century CE was found in Jordan, but the technique seems to have been lost in Rome between the second and the third century; it then reemerged in Syria between the eighth and the tenth century. In the Middle Ages glass began to be more widely used, as stained-glass windows in Gothic cathedrals testify, yet it remained extremely rare until the late fifteenth century, when Venetian skill and trade networks combined to give it a permanent place in our world.[27] Similarly, the glazed brick formula and mixing techniques of the Babylonians were completely lost for more than 1,700 years. Though partially rediscovered in the Middle Ages, glazed brickwork did not come to full flower again until the nineteenth century, nearly 2,500 years after its initial invention.[28]

Even seemingly simple engineering principles can be lost when the chain of human iterative knowledge is broken. The geometric strength of the diagonal, understood and widely utilized by the Romans (Apollodorus first used it on a bridge over the Danube), was lost, as demonstrated by the lack of its use in medieval timber framing. Palladio finally reintroduced it in his bridge designs in the late Renaissance. Even then, the structural theory underpinning the strength of the diagonal, known as Stevin's "triangle of forces," was not proved until 1847.[29]

Practical Applications

In his reportage on the Cluniac revival of monastic life in the twelfth century, G. G. Coulton observes that engineering problems of the resultant building boom "taxed the builder's resources to the utmost, and, often, even beyond.... An imperfectly civilized people was gradually learning, by a path of gropings and failures and half-successes."[30] Some strategies failed in short order. Many of these trial-and-error "improvements" turned deadly when collapse and catastrophe ended the lives and dreams of the builders who created them. A scaffolding collapse at Canterbury Cathedral in September 1178, for example, left master mason William of Sens gravely injured, and for a long time after, he attempted to direct the work from his bed. When his injuries failed to improve, he was forced to surrender his commission.[31] Another well-known incident occurred in Troyes, France, in 1530, when master mason Colin Millet was killed by the fall of a flying buttress that had failed after its centering was removed.[32] Historical records, in fact, brim with notations of people killed, crushed, and maimed by falling stone and other masonry mishaps. The registers of Saint André in France record "a mason's labourer who had spoiled and broken his shoulders at the work of the said church; they gave him 3 sols tournois (3 francs) for charity to the poor man." Or, "The lime burner's servant who had fallen into the lime-pit; they gave him 3 francs of Bordeaux 'for the love of God.'"[33]

Pushing the limits of what has been done before has always been the essence of the iterative architectural process. The multiple collapses at Beauvais Cathedral in France are instructive in this regard. As the tallest, most capacious and audacious of the European cathedral projects, it was also one of the most ill fated. Begun in 1247, the apse and choir were completed in 1272, only to collapse twelve years later. The structure was begun again and completed to the same height, but this time with twice the buttressing and only a third of the original length. Halted during the Hundred Years' War with England, building got underway again in the sixteenth century. Between 1564 and 1569, the tower was constructed, finishing the major work. At 510 feet in height, it was then the tallest masonry structure anywhere in the world. Four years later, it also collapsed.[34] Whether out of wisdom or fear, the cathedral spire was never rebuilt. Had it survived, Beauvais Cathedral would have remained the tallest built structure until the twentieth century, when the Singer Building in New York was completed in 1908.

Historical records are replete with strange and sometimes grotesque latent failures of masonry. One difficulty for master masons lay in drawing conclusions from structural collapses, because the failures so often seemed inexplicable. Buildings or bridges could stand for long periods—months or years—before failing for unaccountable reasons. The infamous London Bridge, with its nineteen arches spanning the Thames, failed so many times in its six-hundred-year lifetime that by the sixteenth century it inspired the children's song still sung today.[35] One of the most famous buildings in history, the Hagia Sophia, was originally completed in 537 CE, only to collapse thirty years later. The design was modified and the structure rebuilt.[36]

Social and cultural habits may certainly be implicated in the periodic collapse of medieval buildings. Unlike much Egyptian and Roman construction work, which employed a vast number of slaves as forced labor, medieval craftsmen were usually professionals, directed and paid by guild-initiated masters. Even during the great period of English civil construction under Edward I (r. 1272-1307 CE), when able-bodied men were summarily rounded up to labor under forced impressment, wages were paid and the workforce was recruited from experienced craftsmen.[37] Charged with carrying forward any number of underfunded projects, they were often overwhelmed, and overreached their abilities. In any case, eleventh- and twelfth-century cathedrals were built with much smaller work crews over much longer periods and with greater economy of materials. This may well have increased their propensity to fail.[38]

Today, catastrophic failures are decidedly rarer. Yet our understanding of the loading from wind and thermal movement remains imperfect. Indeed, in May 2004, the masonry vault in the newly opened Air France terminal at Charles de Gaulle Airport in Paris collapsed upon stunned passengers a mere eleven months after opening. Four people were killed. To the global humiliation of the French engineers and contractors, subsequent reports suggested that the vault was not engineered to support its own weight, and consequently its piers eventually penetrated the vault itself.[39]

It is not hard to imagine the bewilderment of premodern people in response to similar shocking architectural failures. Although medieval naturalists—the precursors to contemporary scientists—worked with great seriousness, in their deeply religious and/or superstitious societies, supernatural forces were readily invoked to prevent catastrophe. Coulton reminds us that while building a church, if some portion collapsed, the damage was surely deemed less than could otherwise have been expected thanks to "the merits of the particular saint concerned."[40] In such an environment, when the construction of the building itself is a quasi-mystical, mysterious, even mythological phenomenon, a material collapse is even more so, inspiring awe and terror as punishment from God, and a manifestation of his ultimate power. What God has a hand in making, God can also have a hand in destroying. In Durham, England, Bishop Pudsey (1154–1198 CE) began to build a Lady chapel at the east end of the cathedral when, according to the contemporaneous historian Geoffrey de Coldingham, "the walls having been erected to scarcely any height, the work at length yawned in fissures and gave manifest indication that it was not acceptable to God."[41]

Although these value systems considerably slowed the rational analysis of cause and effect, they also served to reinforce the "sacredness" of the empirical knowledge held by the guild.

When God's hand is evident in the building (and destruction), the very fundamentals of the structure become sacred. This seemed to inspire the workforce as well as the community in supporting the building effort, since being closer to it, seeing it, being in it, was thought to bring one closer to God. Sometimes the temple or cathedral had doors of gold or silver. When these shone brilliantly in the sun, one couldn't even look directly at the structure without having one's eyes burned by the power of God, the saint, or whatever sacred entity pertained.[42]

In whatever way the blame for a failure may have been worked out, the safety of stone structures has always been a chief concern of building codes and the primary obligation of the builder. Perhaps the earliest example of such a code appears on a stone column, now at the Louvre in Paris, dating from the time of Hammurabi, the king of Babylon in the eighteenth century BCE. A section of Hammurabi's code proclaims, "If a builder has built a house for a man and his work is not strong, and if the house he has built falls in and kills the householder, that builder shall be slain."[43]

It is not widely understood that structural building codes x prescribe materials and methods only from the exclusive perspective of building and material safety. The codes never concern themselves with issues of longevity, craftsmanship, or aesthetics. While building codes may articulate a method of attachment or installation that is considered safe, they arrive at this assessment often by ignoring better, more thoughtful ways that materials *should* be attached or installed—methods that both look better and ensure the longevity of the structure. The unfortunate result is that poor installation techniques are too often advertised as "code compliant" when in fact the code is merely a prescription for safety, a low bar that must be hurdled, not an alibi for poor details that may fail over the longer term. A case in point is the modern building code that allows stone veneer to be backfilled against wood sheathing with only a paper barrier. Without an air space behind the veneer and adequate water drainage, many applications have failed as condensation from the shifting dew point enters the building envelope, rotting away the shear wall. Significantly improved methods are well known and documented.[44]

As I hope to have made clear, building codes, ancient and modern, do not necessarily reflect best practices. And as building masters retire and building knowledge changes, trade secrets that for centuries have been passed faithfully from generation to generation can easily become disconnected from their core principles. The French master mason Jean Mignot, for example, when brought in to consult on and finish the Cathedral of Milan in the year 1399, struggled with this very issue. The historian and British civil engineer Jacques Heyman explains that "Mignot's victory, and subsequent appointment as architect at Milan, did not mean that he himself understood the rules to which he worked; rules which were, to Mignot, something out of a book, and which had been tested by time and practice but whose meanings and indeed, whose very reasons for existence, were becoming more and more dim with the passage of time."[45]

This problem has even occasionally occurred in my own businesses, usually when procedures were blindly followed because they were rules. Then and now, the underlying principles of application and technique from which the Rules of stonework spring are increasingly hard to transfer and teach from factory to factory, team to team. In medieval times, the knowledge transfer from master to apprentice exacerbated the problem. An apprentice was ordinarily given only essential rules to guide his work, with the underlying principles not articulated until the last moments of knowledge transfer. As Schwaller de Lubicz notes, "The ancients never 'popularized' anything; to the uninitiated they provided only the minimal *useful* teaching. The explanation, the philosophy, the secret connection between the myth and the sciences were the prerogative of a handful of specially instructed men."[46]

Master mason Mignot's famous quote, *ars sine scientia nihil est* (art without science is nothing), reminds us of the delicate balance between theory and practice.[47] It is easy today, in an age of advanced engineering standards and formulas, to fault ancient buildings and builders for lacking any coherent theory, but that ignores what is so obviously true about them: their structures have survived the test of time. Even faulty theory, balanced with good practice, can produce spectacular results. No credible voice can claim that ancient structures disobey the rules of modern theory or that they stand in spite of our current understanding of sound building principles. Rather, we marvel at the quality of empirical knowledge that managed to produce such buildings without modern science. Mignot and his colleagues in late medieval Milan may have made some mistakes in their calculations that would not occur today (French airports aside), yet they went on to produce a cathedral of such consummate skill and human expression that in modern times we are unlikely to match it.[48]

Our Modern Reality

Today, we find that ancient building standards have been reversed: theories, codes, and mathematical models abound, while the practice and understanding of technique have fallen to a historic low. The ability to computer-model stresses in a structure has not given rise to more spectacular masonry. Rather, it has in large part turned us away from stone's natural variability toward human-made materials with more consistent uniformity. Modern stone practices, then, have become casualties of the superiority of scientific theory, and this has come at considerable cost to expressive stonework. In this context, the ancient and hard-won Rules can themselves appear as intellectual trifles. We may feel an affinity for them and the integrity of what they offer, yet remain disconnected.

The Rules resonate most cogently within the practice of stonework itself. The empirical knowledge that fashioned the great buildings of the past grew out of an understanding of the innate strength of the material, and from a crucial, ongoing search for human safety and protection. The psychologist Les Parrott III reminds us that the "human brain is hardwired in a unique pattern to seek this safety."[49] On an unconscious level, humans are always assessing the safety of stonework. The Rules of Bondwork guide us in this task by alerting us visually to that which would harm us or be unsafe. Our first clues that a particular wall is contradicting one of the Rules will be somatic. The image of the work sits uncomfortably in our bodies, with our minds catching up eventually.[50] This is why modern applications that ignore visual safety undermine stone as a primary building material, since they work against our unwritten expectations. Knowing the Rules of Bondwork helps us avoid this error by codifying primary techniques, practices, and uses of material that support the safety human beings are wired to seek.[51]

In the twentieth century, the guild system finally broke down. Transmission of the sacred knowledge in Europe and the United States was disrupted utterly by the World Wars. Empirical knowledge and technique were abandoned in favor of theory and mechanics, and much of the poetry of stone drifted away. Brunés puts it brutally: "Of the knowledge which had at one time been the bloom of all wisdom only the sad remains were left in the form of craftsmen's guiding rules."[52]

As we shall see, a good deal of the stonework assembled today confounds basic, intuitive guidelines that have defined masonry from the beginning. How exactly did we lose our way?

The reasons are complex and worth investigating. Certainly, one factor to consider is the massive change in the quantity of our urban architecture. We may have become numb to our perceptions of the built environment by the sheer abundance of impressions we now absorb. Consider how our ability to travel nationally and internationally has created a greatly expanded visual repertoire of architectural possibilities. Even locally, in one daily commute, an ordinary person may visually ingest more architecture than our ancient peers encountered in a year, a decade, or a lifetime. By way of example, imagine the profound changes to the landscape that have occurred in just the last few generations. Interstate freeways, power lines, prefabricated metal workshops, strip malls, mobile homes, flat digital signs, and round hay bales are but a few of the new forms added to our visual horizon during the last sixty years. These changes alter the look and feel of city and country in profound ways. They have come to define the built environment of our modern world, and virtually all of them can be traced to the creation and adoption of new technologies.

To explore one new technology affecting masonry, consider elastomeric form liners. These offer a simple method that allows the imprint of almost any texture to be cast repeatedly, and with precision, into concrete. More and more frequently, those imprints approximate stone patterns. Entrepreneurs can take impressions from existing stone walls or, more commonly, make up their own. City councils and state and federal transportation agencies now mandate that these products be deployed along our nation's roads as noise barriers. A drive down almost any highway in almost any urban center in the United States brings a jumble of stonelike impressions. These ape the basic pattern and texture of stone, creating an exhausting repetition and proportion that numb the eye and dull the senses. The predictable result is manifested in the trouble we face determining what is visually correct, harmonious, or safe.

For most of us, some change in cognitive visual reasoning has already taken place, perhaps a measure of how so much poorly conceived work has already entered our world, dulling our perception and expanding our tolerance for the unexpressive and ugly. Our ability to hear the small voice of inner instinct has been drowned out by the visual eclecticism of modern urban centers. In this greatly expanded visual palette, our innate, intuitive understanding of the Rules is far less available. Counter to what one might expect, the more we see, the less certain we are about what our instincts are telling us. The more visually numbing and shoddy work we encounter, be it elastomeric stone impressions or pastiche architecture, the more these images corrupt our ability to discern visual safety.

Another way to explore this cognitive change is through the prism of evolving architectural invention, especially if we can monitor our own bodily response as this invention is considered. The iterative process can be viewed as an engine of change that encourages and rewards fresh solutions. In this light, it may be instructive to imagine the awe that was inspired when one of the earliest stone lintels was raised into position at the Lion Gate of Mycenae, Greece, in 1250 BCE, creating a twenty-five-ton roof to the entrance portal. This lintel measures sixteen and a half feet in length over an opening spanning nine feet, making it the largest lintel in the ancient world.[53] The visceral experience of walking under that hoisted stone would have created an unparalleled emotional impact.[54] Today we have grown accustomed to dramatic entry sequences that were lifted into position by hydraulic cranes, but in the ancient world, such invention and accomplishment pushed the boundaries of what had been done before. By successfully leveraging the core attributes of the material—strength through mass—the ancient Greek builders created something entirely new, evolving the use of stone and, indeed, architecture itself. Humans, having encountered visually strong stone lintels for now thousands of years, have come to recognize them as embodiments of safety. Looking at stone architecture today, we may feel uncomfortable and notice something amiss if the strong lintel is missing.

Some modern changes that have demeaned our aesthetic sense have more in common with ill-informed fads than historical inventions. For example, within my own guild in Siena, it was well understood that entasis—the bending of the top two-thirds of a column to counter the illusion that vertical lines of the column are concave—was to be used only on columns over fourteen feet in height.[55] This makes perfect sense in the context of the height of the human eye as it gazes upward at a tall column. Yet no minimum height for entasis has been seriously articulated in hundreds of years. In fact, the bowed column has become such a popular feature that even wooden columns on furniture routinely show entasis, and columns without entasis are now thought to be primitive. This detail, once architectural, has in the present day become completely unmoored from its original reason for being, instead evolving into its own strange

convention. Such an example may seem harmless enough and might arguably be cast as an iterative stylistic improvement. But what is it really, beyond fad and fashion?

Without the imperative of tying safety, purpose, and emotional well-being to shape and pattern, these pointless stylistic changes proliferate, quite disconnected from function. When our adaptive nature integrates, accepts, and even takes for granted the patterns and forms that undermine our sense of safety, we pay a high price for dulled perceptions. We are already paying dearly when we elevate the visually banal as an expedient way to decorate our built environment.

The old adage tells us that history is written by the victors. Just so, our architectural heritage is, in a sense, written by the successes that have survived from our past.[56] Obviously, we base much of our current knowledge on what is still standing. As the architectural scholar William Richard Lethaby, writing at the turn of the last century, observed, "A notable building, indeed any work of art, is not the product of an act of design by some individual genius, it is the outcome of ages of experiment."[57] Moving the advancement of design and technique forward with incremental improvements has always been one of humanity's most basic impulses and finest accomplishments. This continual evolution is part of a natural human urge that drives us to mine the successes of the past for future inspiration.

In this way, our return to the road map of the Rules of Bondwork is timely, since throughout human history we retrace our steps when we discover that we are lost. Such a time is again upon us, for modern usage of stone has stripped bare its expressive potential. The Rules operate as bedrock principles with which we navigate, diverging only with great caution. As we have seen, powerful new tooling allows us to slice and dice stone, yet to build something meaningful, something touched with the emotion and reach of human expression, deeper processes must be involved. The Rules can guide us in this reconsideration. Given their history and importance, it is time now to turn to the Rules of Bondwork themselves.

OPPOSITE
The twenty-five-ton lintel at the Lion Gate of Mycenae would have inspired awe and wonder among those entering the ancient citadel. Both the method and the rudimentary strength calculations of the assembly would have been empirical, and closely held as sacred knowledge by the builders. Mycenae, Greece

II.

THE RULES OF BONDWORK

Chapter Three

FOLLOW THE GRAIN

Often in architecture, the material employed imposes its own laws.

—ALFONSO ACOCELLA[1]

The Rules of Bondwork fall into three main categories, all supporting the safe assembly of human expression through stone. The first relates to making sure the material succeeds. The second concerns itself with making sure the wider assembly is safe and strong. The third focuses on how we might wring as much expression as possible from the effort. Some Rules appear to follow simple common sense, but as the history of surviving structures might testify, this has not always been the case. While the Rules themselves may appear conceptually simple, their application can often prove difficult and complex.

The photographs illustrating the Rules will help decipher most of the stonework that one might encounter, whether locally or in remote lands. But for those with the persistence to study the technical explanations that follow, I hope to reward that effort with a glimpse into the elusive balance between practice and theory. As it turns out, the Rules of Bondwork are not blanket prescriptions to be obeyed in all circumstances or to be applied witlessly to every complicated design issue. They certainly were not intended as the final gauge of quality for the great variety of human expression in stone. However, they do form the skeleton on which most masonry architecture in our history has been hung, and they serve as the technical basis for masonry practice itself. Though straightforward enough, the Rules have a technical edge because they grew out of technical issues, technical failures, and technical glories.

Lay All Stones in Their Natural Bed

When masons speak of "bedding the stone," they are referring to setting stone horizontally, with its primary grain orientation or bedding plane parallel to the earth. For many readers, it may come as a revelation that building stone, much like wood, has a grain structure that is critical to its use. The primary grain follows the original layers (igneous or sedimentary) created when the material was first formed geologically. Originally these layers paralleled the earth's surface, although we may today find them resting at nearly any angle. As we all know, the Earth is constantly moving through a variety of natural events. What geologists call "mountain building events" have radically disrupted the original horizontal layering of the Earth's crust. These

profound changes are called orogeny. The word comes to us from the ancient Greek and literally means "mountain creation or origin." With careful observation, we see clear evidence of disrupted layering, or orogenesis, almost everywhere in the natural environment. In sedimentary stone, the horizontal layers are more obvious to the eye, since differing minerals color the sediment as grains of sand, mud, silt, or organic matter accumulate over time. These layers become the migration path of groundwater that carries the minerals and cementing agents, which, under pressure, heat, and time, turn the disparate grains into a cohesive and worthy building material. That same groundwater also brings color into the stone as the mineral constituents migrate with it, bonding and coalescing.

It is the slow cooling of igneous rock over many tens of thousands (if not hundreds of thousands) of years that serves to organize its molecular structure into horizontal layers.[2] This cooling action not only creates contraction cracks or "joints," as they are known, but it also aligns the developing crystal structure prismatically, causing planes of weakness.[3] Noted for his books illuminating geology, the writer John McPhee explains the process in simple terms: "As the magma crystallized and turned solid, certain iron minerals within it lined themselves up like compasses, pointing toward the magnetic pole."[4] The resulting alignment created a plane of weakness called the primary grain. The primary grain originally ran parallel to the Earth's surface, no matter how deep the deposit formed. This plane of weakness in stone is called the rift and is of "considerable scientific interest and of much economic importance. It is an obscure microscopic foliation...along which the rock splits."[5] The earliest mention of rift in the geological literature dates from 1778, when it was noticed that granite millstones cut with their largest diameter parallel to the rift "were much more readily worn than those cut at right angles to it."[6]

When we try to follow the grain of stone in its natural setting, we must keep in mind that the original orientation of the grain may have changed dramatically as geologic mountain-building events disrupted and reordered its context. In whatever position we currently find the primary grain, building stone is properly quarried when it is extracted in its *original* relationship to the Earth—those horizontal layers created when the material was laid down (sedimentary stone) or cooled (igneous stone). Although some materials (such as slate) may even be quarried when the bedding plane (now) runs vertically, most building stone is never

OPPOSITE
The horizontal grain of the Egyptian sandstone used in the Temple of the Nobles is instantly apparent, showing that the material was properly set in its natural bed. Setting material horizontally increases its bearing capacity by a factor of two to five thanks to leveraging the innate strength of its inner grain. Aswan, Egypt

TOP
Seashores often reveal the original bedding planes of rock. Viewing the earth in section, we bear witness to the often-dramatic disruption of these layers by prior mountain-building events. Cannon Beach, Oregon

BOTTOM
In sedimentary rock like sandstone and limestone, the primary grain is often easily identifiable. Groundwater follows this natural plane of weakness, dispersing trace minerals that color the different strata of the rock. The red color shown here is often associated with dispersed iron minerals. Wadi Rum, Jordan

The horizontal layering of the Earth's surface is key to understanding the grain of stone, since the primary bedding plane (and the plane of greatest weakness) mirrors that of the original orientation of the Earth's layers. Agdz, Morocco

quarried when the bedding plane exceeds thirty degrees from the horizon.[7] An abandoned quarry on Easter Island hints at the relationship between bedding planes and the technical difficulty to be managed when removing stone from the earth. The architect Vincent Lee explains that on Easter Island, in the volcanic crater called Rano Raraku, the largest moai ever carved can be found. El Gigante, as it is known, would have weighed an estimated 270 tons. The sculpture is "lying on his back, head uphill, still attached to the sloping bed matching the angle of the underlying rock strata, about 24.5 degrees (55 percent) above the horizontal."[8]

The current resting angle of the Earth's stone strata is called the "dip" in quarry terminology, and in geology as the "angle of repose." The usage derives from the mechanics of quarrying, which is more concerned with the angle at which a block can rest on the sloped surface of the quarry floor without sliding.[9] Quarrying with the primary grain takes advantage of the natural planes of weakness in the stone, helping to ease its separation from the larger mass of the Earth. By accessing the primary grain, we exploit the weakest plane of its mineral composition and avoid the strongest orientation of its constituent elements. Relatively small amounts of force applied in a line along this plane of weakness can cause the stone to split apart.

Learning about this plane of weakness, one might be tempted to think that the work of quarrying stone—breaking out blocks of usable dimensions—would be relatively easy. Simply follow the

LEFT
This quartzitic sandstone is being quarried in the traditional method with simple tools: hammers, chisels, and iron wedges. Electric or pneumatic drilling and the use of "plug and feathers" instead of wedges changed this ancient technique only slightly at the dawn of the Industrial Revolution. Neijiang, China

RIGHT
Once a stone block has been successfully extracted, it must be moved down the mountain and loaded onto transport. In most countries, it goes first by truck and then by rail or ship to its final destination. Songjiacun, China

primary grain, right? But in practice, beyond the huge weights involved, beyond the imperfect detective work in deciding exactly where to quarry, the work itself is highly complex. I have often exclaimed, "A worthy opponent!" after a particularly thorny week of work in a quarry. One cautionary tale, as related by W. Lewis Barlow of the US National Park Service, concerns the quarrying of the granite blocks for the Bunker Hill Monument in Boston in 1826. The stone was quarried by an architect who quit his practice to "quarry the stone himself" after rejecting the quoted cost of 90 cents per cubic foot. Certain that he could provide the material for a mere 20 cents per cubic foot, he spent several years trying to solve the quarry problem by creating the first primitive railroad in the United States, on which stone was pulled along metal rails by animals. Unfortunately, the total cost of the stone continued to be "much higher than the original quote."[10]

In practice, following the first Rule of Bondwork—"lay all stones in their natural bed"—is even more complicated, since there are varying degrees of strength between primary, secondary, and tertiary grain structures. Understanding these differences is key, since the core premise is to keep the primary grain of stone in relation to its highest strength. Honoring the material's relationship to the Earth and gravity makes perfect sense on an intuitive level, but it also turns out to be good science. Modern material testing has shown that stone is typically two to five times stronger in compression perpendicular to its primary grain than perpendicular to its secondary grain. Because of this, stone engineering requires that the bedding plane be clearly indicated on all laboratory samples for both flexural and compressive strength analysis.[11]

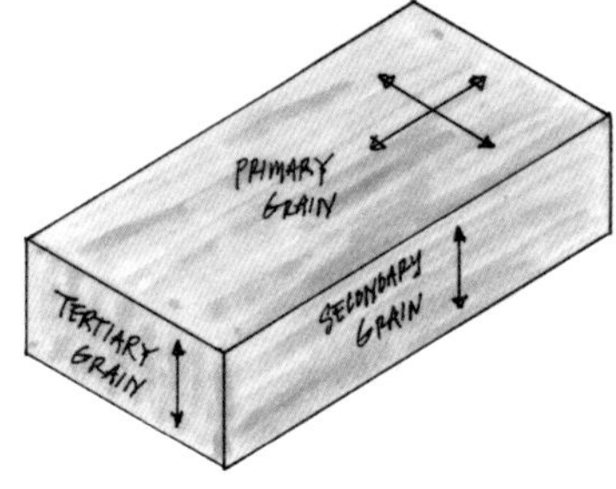

In a standard quarry block, the largest face will always be along the primary grain so as to exploit this natural plane of weakness. By the same logic, the smallest face of the block is always the tertiary or end grain, offering the most resistance to breaking free. Thus, the very shape of the final block is determined by the varying resistances of the respective grains.

Not all stone has directional strengths and weaknesses, and the famous statuary marble of Carrara, Italy, is a case in point. Prized by sculptors since Roman times, the stone is considered isotropic, or independent of grain.[12] Originally a sedimentary limestone (all limestones are sedimentary by definition and have strong grain directionality), the material was subjected to intense amounts of heat and pressure, converting it into marble in a metamorphic process over hundreds of millions of years.[13] Marble and limestone are chemically the same material; the difference lies in marble's loss of grain structure. The result is called "freestone," and it is isotropic rather than anisotropic (having grain structure). Of course, nothing in geology is quite that simple; there are even rare limestones that have also lost their grain structure due to geological events. They, too, have somehow become isotropic without fully converting to marble. The famed Portland stones of south England, widely used by the seventeenth-century British architect Inigo Jones, are considered freestone. In the vernacular of the time, a stonemason and quarry broker describing Portland stone to a client in Amsterdam wrote, "Lay this flat wyes side ways sett itt on end all is alike and as durable one way as the other."[14] Essentially, the smaller the crystalline structure within the stone and the more homogeneous it is, the more highly prized. In granite, stone with the smallest crystals is considered "statuary grade" and typically reserved for sculpture. To a carver, freestone is valuable, since the resistance to the chisel is even throughout. The stone lacks a preferential direction.[15]

G. G. Coulton suggests that the term "freestone" informs the nomenclature of Freemasonry, a freemason being a worker in freestone. He notes that the term "freestone" predates "freemason" and that building accounts distinguish between "freemason" and "rough-mason" or "hard-hewer," "very much as our fathers distinguished between 'whitesmith' and 'blacksmith,' and as the Germans still call the latter *Grobschmied*, 'rough-smith.'"[16] There are, naturally, dissenting viewpoints. The well-respected Gerhard Rosenberg writes, "It is highly unlikely that the real free masons (a name that describes a mason who has been declared free from the bonds of apprenticeship and accepted in the community) had anything to do with the philosophical outlook of Freemasonry as we understand it."[17] My conclusion is that freestone carving gave the name to freemasons, who traveled as initiated free men and assembled the earliest records of the medieval guilds. These free citizens were not bound to the land or feudal lords by birth, but were free to own property and travel as their work required, a lucky coincidence.

When discussing the grain of stone, we can again follow the analogy of wood. The moment we pick up a board of any length and see it sag in a particular direction, we understand how the grain determines its strength. Further, we see how the rectangular shape of the milled wood mirrors and supports the relative strengths of the differing grain directions. Properly milled wood will always have the primary grain running in the widest dimension, precisely as a stone block will. In wood furniture, the stress of use thus becomes aligned with the strength and weakness of the material. Perhaps the Nobel laureate scientist George Wald said it best when discussing the masterful furniture of George Nakashima: "Each flitch, each board, each plank can have only one ideal use. The woodworker, applying a thousand skills, must find that ideal use and then shape the wood to realize its true potential. The result is our ultimate object, plain and simple."[18] Indeed, we are not likely to build anything of substance with the grain of the wood oriented in the wrong direction. The more obvious flexibility of wood makes proper grain orientation apparent within a very short time of working the material, and should we mistakenly orient our lumber in the wrong direction, the wood will quickly fail under load. Building in the wrong orientation with stone can be equally disastrous but harder to diagnose. While stone may be less likely to crumble away in the short term, it is very common for improperly oriented stone to delaminate and fail by degrees over many decades.[19] The Australian writer Paul Hasluck said it best: "Build with the cards on end, and it will be easy to peel them off one by one; but place them on their backs, or on their faces, and such disintegration is practically impossible."[20]

The grain of stone has been likened to building with a pack of playing cards. The Australian writer Paul N. Hasluck said it best: "Build with the cards on end, and it will be easy to peel them off one by one; but place them on their backs, or on their faces, and such disintegration is practically impossible." [Paul N. Hasluck, Building Stones (London: Cassell, 1904), 31.] Boston, Massachusetts

In the northeastern United States, it is common to see this type of failure in brownstone buildings. The face-bedded material—stone laid up in the wrong grain orientation—delaminates and disintegrates. As the Rule of Bondwork illuminates, such troubles might have been easily prevented. This first Rule concerning grain was well understood by initiated craftspeople, some of whom predicted disaster early on when brownstone buildings were constructed. Correspondence found in the archives of the Cathedral of the Immaculate Conception in Albany, New York, for example, gave clear warnings to the church authorities overseeing the work. One letter warned ominously of "problems associated with the quality of the building stone."[21] Unfortunately, the work proceeded. Then, 150 years later, the building faced catastrophic failure of the stone veneer, and large pieces delaminated and fell onto a pedestrian walkway below. To this day, because of insufficient funding, only the stone on one spire and the nave and transept clerestories has been replaced; the rest of the building remains shrouded in black safety netting.

Brownstone failures can be traced to low budgets, short-sighted economics, and ignorance of the Rules of Bondwork. While the building boom of the late nineteenth and early twentieth centuries employed highly skilled stoneworkers in granite and limestone quarries, low-skilled laborers arriving from failing economies in Europe were taking up lower-paid positions in sandstone quarries. The difference between these two groups was significant. The most skilled, newly arrived stoneworkers were guild-initiated, fluent in the Rules of Bondwork. Many found immediate employment in the classical stones of marble, limestone, and granite. Those with lesser skills and training worked with sandstone, imitating the shape—if not the substance—produced by initiated craftspeople. Today, anyone driving through neighborhoods of crumbling brownstone structures and unaware of the skill differential between initiated craftspeople and common laborers might conclude that the fault lies in the material itself. Are the tens of thousands of crumbling brownstone structures simply proof of a material incapable of surviving the brutal freeze-thaw cycles of a northern climate? As it turns out, no. Properly quarried, most sandstones—and brownstone in particular—are excellent building materials for almost any climate.

A review of the buildings of the nineteenth-century architect Henry Hobson Richardson, whose masons followed the first Rule to the letter, shows the expressive potential of brownstone at its finest. None of Richardson's brownstone buildings are failing structurally, because he employed guild-initiated craftspeople who understood how to translate architectural design into the medium of stone as well as aligning that expression with the highest and best use of the material itself. Richardson's long and important relationship to the Norcross Brothers—his builders, who "could only be described as collaborators" given their experience in stone and ownership of quarries—clearly influenced the direction of his architecture.[22]

Although the prohibition against face bedding was a tightly held "secret" of the guild, it was also knowledge recorded and shared. Inigo Jones, privy to such trade secrets, recorded that "stones should be cut across the grain to avoid face bedding." This note to himself was found in the margins of his well-worn copy of the Italian translation of *De architectura* by Marcus Vitruvius Pollio (known today simply as Vitruvius, writing in 30–20 BCE), a military architect and engineer who had worked for Julius Caesar.[23]

One trade secret—particularly relevant to strong-grained sandstones like brownstone—involves the curing of the stone prior to working. When blocks of stone are first quarried, they are still heavy with groundwater, or "quarry sap." In addition, freshly quarried material has been "de-weighted" from the intense geological pressures within the contiguous rock as well as the accumulated weight of unconsolidated rock that had been pressing down upon it.[24] Part of the curing process allows the slow resolution of the forces within the stone after de-weighting. Without this step, sandstones and some limestones are prone to delamination once the downward pressure of the layers atop them are removed. Although the time needed for the stone to resolve itself varies from quarry to quarry, it is not uncommon to let blocks "rest" for ten to twenty-five weeks after extraction.

Although not technically alive, freshly quarried stone feels "lively." It's a curiosity that is reflected in our language. Today we speak of the "living rock" when we sense groundwater still moving through the material. Much like a newly felled tree, freshly quarried stone can be worked in a number of ways that will not be possible once the stone dries out. Many of us who were born and raised in the United States—a wood culture—will be far more familiar with this phenomenon from framing timber. In rough carpentry, driving a 16d (standard) nail into a newly milled two-by-four is a satisfying experience. But should that same lumber dry out—the sap of the tree no longer present—the nail can barely be driven through the mass. The lack of sap has fundamentally altered the wood's hardness and workability.

Stone is much the same. In slate, for example, the fine quarter-inch shingles that grace the roofs of our best stone buildings are split by hand within the first year after quarrying, when the stone is "fresh," or still flush with natural quarry sap. Since most US slate comes from the northern quarries of Vermont and upstate New York, where freezing temperatures make winter extraction impossible, slate is typically quarried only in the summer months. Once the blocks are free of the earth, they are often wrapped in burlap and stored in shaded quarry sheds to delay the curing of the stone. Once freezing temperatures arrive each winter, quarrying stops and the work moves indoors. The stored block inventory is now split by hand into rough sheets, ideally a quarter inch in thickness, that are then trimmed into stone shingles or "slates." It may be surprising to learn that any unsplit blocks of the season will be abandoned and considered useless, since once the block is dry or fully cured, it becomes intractable, its primary grain no longer able to be cleaved. Masons and stone sculptors call this "cooked" or "dead" stone.[25]

For the same reason, commercial carving is done only on freshly quarried blocks—typically out of the ground for less than a year—since once the block has seasoned, it becomes too hard to carve.[26] This is particularly true of harder stones like granite. One needs every advantage to make the difficult work economically viable. Some carvers have been known to store their freshly quarried stone in rivers, until the slabs are needed, to extend the capacities of living rock a little longer. And quarried blocks may be preserved indefinitely by burying them after quarrying and letting the groundwater continue to flow through them. Generally, this must be done within three months of breaking them free, and the blocks must be buried below the frost line with the primary grain parallel to the Earth's surface to facilitate the flow of the groundwater. I have used this strategy successfully to allow production through the winter months on projects with tight timelines.

Once a stone block has dried—the groundwater gone—it can never be revived. This is not to say that one cannot collect a random piece of "dead" stone from the surface of the Earth and carve a name into it. One certainly can. But the experience is much like framing a house with several-year-old timber found on the side of the road. The process is hard on the tools and hard on the hands, and ultimately the result will be less successful than if freshly quarried living rock had been sourced.

On the other hand, some stones—such as Bath limestone—are notoriously difficult to dry out and must be fully cured before processing. Blocks of Bath limestone require a full year to cure before they can be used, since if they "were laid in their green condition, they would weather away rapidly."[27] Likewise, with the brownstone of the Northeast, when summer approached, the guildsmen quarrying the stone took great care to cure it. Because they were full of groundwater, the blocks were particularly vulnerable to frost if shipped uncured during winter. Some of this damage would be obvious, and the stone would be discarded. Other blocks, weakened by frost but not obviously failing, would be used, and if installed with the wrong grain orientation, would fail, often in short order.[28] Meanwhile, initiated guildsmen may have smirked in silence, their blocks buried in sand for the winter to slowly cure and leach the quarry sap away, creating stronger stone for their projects. If not "sacred" in the modern use of the term, their critical knowledge was kept secret for the benefit of guild practitioners and their dedicated clients, making an enormous difference in the stability of their architectural projects.

Since groundwater is the key to stone's preservation and resilience, it makes perfect sense that on the tops of mountains we find the crumbling de-evolution of "dead" rock. As mountain-building events raise the Earth toward the heavens to create new mountains, the groundwater no longer reaches the rock of the newly formed peaks. Without this life-giving flow, the stone essentially dies—it dries out, contracts, delaminates, and is split apart by frost and water. McPhee reminds us that "the sea is not all that responds to the moon. Twice a day the solid earth bobs up and down, as much as a foot."[29] That kind of force is more than enough to break hard rock. As mountain peaks crumble, the broken fragments make their slow journey down the newly formed valleys during the epochs that follow. The particles from this degradation create the sedimentary rocks that form the second main group of building stones of which New York brownstone is a member.[30] As a group, these aqueous or sedimentary rocks form the majority of the Earth's surface, by some estimates more than 70 percent.[31] In contrast, only a small portion of the Earth's crust is composed of igneous or metamorphic rock, perhaps as little as 8 percent.

In the same way that the ruins of stone architecture whisper their histories to the patient observer, fine grains of rock washing down our rivers eventually find their way to our beaches, carrying their stories within them as well. For example, rounded sands are far older, their sharp edges worn away in the tumble over eons. It reportedly takes a million years for a grain of sand to travel one hundred miles in a river.[32] Black sands are typically volcanic basalts. Paler sands derive from the breakdown of granite, their colors due to their quartz or feldspar constituents. The prized white sands of travel-brochure beaches come from wave-crumbled coral reefs as digested by parrotfish that ingest the coral polyps (limestone).[33]

Irrespective of the color of the rock grains that make up the sand, when eons of time, pressure, and groundwater exert their influence, those same grains are well down the track of metamorphosis. No longer loose and shifting, the compressed mass starts to behave as a solid. Even the color can change as the mineral constituents of the surrounding earth become soluble and—traveling with the groundwater—penetrate the newly compacted mass. This is particularly true with quartz-based sands, since once minerals permeate the inherent crystalline structure, the color locks in permanently. The deep, rich red of New York brownstone was colored with a diffused iron. In Egypt, an excess of iron also infused the fine local sandstone used to build the Temple of Luxor, but resulted in a color that is more khaki than red. Regardless, there is so much iron in the Egyptian "ferruginous" sandstone that, when near it, a sensitive compass will point to the Temple rather than north, regardless of where

Occasionally, dispersed iron in stone is especially soluble. In the decades after installation, the staining mineral migrates to the surface face, giving it a reddish or rusty cast. Srirangapatna, India

one is standing.[34] It is these minerals that typically "weld" the fine grains of sand together into a building stone. Certainly, the intense pressures of the earth also play their part. Paul N. Hasluck notes, "It is generally considered that the stones which lie deepest are more durable than those nearer the surface, on account of the superincumbent pressure, and the consequent density and hardness in proportion as this exists."[35]

For wrought stonework, a bright line divides anisotropic stones (those with grain), setting apart those with "random" grain structure from those with "preferential" grain structure. It is true, if reductive, to say that only stone with preferential grain structures whose strength exceeds a basic threshold should be used for building.[36] Of the thousands of different kinds of rock, there are only four broad categories that have ever qualified as building stone: granite and a few other igneous rocks, slates and schists, limestones (marble), and sandstones. As mentioned, granite is an igneous stone, meaning it started in a molten form and cooled slowly. Sandstone and limestone are aqueous, meaning their constituents were set down in water or air, settling down in layers over millions of years. This makes them sedimentary. Slate, schist, and marble are metamorphic, meaning that through the action of heat and pressure they have transformed and become materially differentiated from their sedimentary beginnings.[37] Marble is particularly interesting, since it started as simple limestone, and despite its metamorphic transformation, remains chemically the same.

Some significant structures have been built with stone that falls outside these four broad classifications. Laterite is a cemented soil rich in iron and aluminum. Occurring in wide swaths in India, northeast Africa, and Burma, it has been used in building for at least a thousand years. Cut into blocks from below the water table, it dries into a reasonably hard "stone" that can be shaped and worked. It weathers well, surviving the annual monsoons of Southeast Asia. Such blocks form the foundation and structural core of Angkor Wat as well as many temples in the Madhya Pradesh region of India.

It may seem strange that judging the strength of stone might be an issue, since resistance to weather is one of the primary reasons to use stone in the first place. Yet determining the durable hardness of stone has been a concern since earliest times. Theophrastus, writing around 300 BCE, advised care in determining hardness in his treatise *On Stones.* Vitruvius advised caution even more explicitly:

Laterite, a cemented soil rich in iron and aluminum, is an interesting exception to the four main classifications of building stone: granite, limestone (marble), sandstone, and slate. A soft earth when freshly dug from below the water table, it becomes reasonably hard as it dries, ultimately possessing stonelike characteristics. Angkor Wat, Cambodia

When it is time to build, the stones should be extracted two years before, not in winter but in summer, then toss them down and leave them in an open place. Whichever of these stones, in two years, is affected or damaged by weather should be thrown in with the foundations. The other ones that are not damaged by means of the trials of nature will be able to endure building above ground. Not only should these precautions be observed for squared stone but also for *caementiciis structuris*.[38]

In 77 CE, Pliny the Elder also weighed in with his classic *Naturalis Historia*. The engineer Henry J. Cowen points out that so much Roman stonework survives thanks to the careful selection of durable stone.[39] We also know that the master British architect Christopher Wren left Portland stone to weather three years to test its characteristics before he would use it in his buildings, a method recalling Vitruvius.[40] In the nineteenth century, the US architect James Renwick Jr. commissioned accelerated weathering tests on nineteen different stones before selecting Seneca Creek sandstone for the Smithsonian Castle on the Mall in Washington, DC.[41]

Unfortunately, such thoughtful examples are historical exceptions, not the rule. More often, stone is chosen for its color or cost without much consideration of its suitability to the prescribed program. Most of these errors are commonly understood, yet the practice continues unabated. For example, Indiana limestone, a medium-density stone, should never be used for exterior paving. Yet despite the Indiana Limestone Institute's own warnings against this practice, member quarries and fabricators routinely supply the stone for this use. Demonstrating similarly shortsighted thinking, none of the stone industry groups in the United States have authorized standardized testing to determine the suitability of a particular stone for the freezing and thawing cycles common in northern climates. Despite Renwick's success with weathering tests in 1847 (he repeatedly boiled and then dried the stone in sulfate of soda), industry "leaders" have successfully stranded these recommendations in committee, decade after decade, letting their customers assume the risk that their product may fail.

Even today, unconsolidated stone—unsuited and technically unqualified as building stone—is sold on the world market in great quantities. This includes many low-density limestones (Texas, Turkey), volcanic tuffs (Italy, India, Indonesia, Mexico), and poorly formed sandstones (Arizona and the Vosges, France).

That last, used to build the Cathédrale Notre-Dame-de-Strasbourg, is now irreversibly crumbling into dust before our eyes.[42] Without modern testing to guide them, we can look with sympathy toward the ancients and their imperfect understanding of the suitability of their local stone. But what are we to think of these contemporary stone brokers?

In some ways, our cultural obsession with what Samuel Johnson called "hunger for novelty" is squarely to blame. With more than five thousand stones on the world market, vendors, continually on the hunt for strange swirls and uncommon shades of color, invariably push the limits of what they consider a "building stone." In a particularly egregious trend, quarry blocks too weak to even make it to market are sawn near the quarry, each slab carefully laid down on a vacuum bed while epoxy resin is sucked through its pores to strengthen it. Since all epoxies degrade in UV light, resinated slabs will yellow and fade after installation in the unsuspecting client's new home. To make matters worse, since UV light penetrates glass easily, the slab will discolor only where the light can reach it, almost guaranteeing dissatisfaction once the stone has yellowed unevenly with age. Real stone, unaltered by resin or epoxy, is universally colorfast.

Sometimes unscrupulous vendors apply resin to stone to make it appear blacker. One can often spot a resinated slab by the telltale dribbles of resin at the edge. True black granite is geologically rare (and thus expensive) and devilishly hard to quarry and split, resulting in small yields and small size formats. All the grain directions of black granite offer tenacious strength. This is particularly true of coarse-grained granites. The famed black granite found in Sweden and Norway has a quarry yield of 3 percent, which means that 97 percent of the stone extracted there ends up being used for gravel rather than dimensional stone products.[43] This scarcity drives up the price and invites stone marketers to call almost any hard black stone "granite," including gabbro, norite, dolerite, diorite, and basalt.[44] However, most of these fraudulently labeled stones can be etched with lemon juice, demonstrating that they are not granites but contain calcium—the same calcium found in limestone and marble formations. Since calcium is alkaline, and since acids and alkalis react and etch, such mislabeled stones are doomed to fail in acid-rich environments like kitchens. Neither limestone nor marble is typically used for countertops, and it is a mistake to use false granites as well. Far more food contains acid than we may realize. Most people know that citrus stains limestone, but few understand that the acids in coffee, wine, and even milk also react to the alkaline calcium. The resulting stain is actually an acid burn, an etching really, and it is permanent, such that no amount of scrubbing will remove it. Rather, it must be mechanically removed by sanding before the original finish may be re-created.

Still, we know that the Romans used marble for countertops and sinks. To prevent staining, they preemptively applied olive oil to create a barrier to food acids. The same may be done today, though food-grade mineral oil is often used instead of olive oil, since it doesn't turn rancid. Mineral oil, used to maintain the finish on wooden butcher-block tables and sometimes taken as a laxative, is completely food safe. But beyond the use of natural mineral oil, no chemical sealer can prevent alkaline stones from being etched by acids. If there were one, we would bathe the Parthenon (made of marble and therefore alkaline) in this magical solution to prevent its further degradation from the sulfuric exhaust of power plants and trucks now permeating its urban environment.

The damage that acid does to limestone is hard to overstate. Consider the cathedral of Notre Dame de Rouen, which sits in a valley of the Seine River where, as recently as 1991, sixty factories belched out one hundred tons of sulfur a day. This sulfuric acid clings to the "delicate stonework, forming a black stain that gradually spreads and penetrates deeper into the stone, exploding it.... What were once distinct noses and eyes on its statues of biblical figures have become puddles of featureless stone. Scaffolding rings the portals to catch falling pieces."[45]

By contrast, true granites are immune to chemical assault and therefore ideally suited to areas of heavy-duty use. True granites are even immune to so-called salt attack, the effect of airborne salts in environments close to bodies of salt water, where salt can be absorbed into building stone and form powerful new crystals under the surface. In weaker stones, expanding salt crystals force the stone apart in a surprisingly short time. Road salts are a common cause of the same problem, and many small communities around the northeastern United States are returning to granite curbstones. Although more expensive to purchase and install, what is termed their "life-cycle cost" is much lower compared to concrete curbs, which can be destroyed by salt attack in as little as ten years. Old granite bridges weather frozen winters without degradation, while newly constructed concrete spans fail with alarming speed in areas with salted roads. The

OPPOSITE TOP
The earliest stone bridges borrow from the simplest post-and-lintel structures. In this example from eleventh-century China, the piers are designed as stone boats to cut through the tidal flow. Granite planks as long as twenty feet span the gap. The bridge is almost four thousand feet long in total. Quanzhou, China

OPPOSITE BOTTOM
It is not uncommon for stone bridges to outlast their modern concrete equivalents by thousands of years. Concrete bridges contain internal steel (rebar) that invariably corrodes and degrades, causing "rust jacking." Leshan, China

ABOVE
The deep black color of this unusual basalt offers a striking, if rare, example of wrought basalt in a coursed ashlar pattern. The color of stone will never fade. Melbourne, Australia

added salt successfully hastens the melting of ice, but it migrates with unfrozen water into the shrinkage cracks of the concrete and attacks the encased steel rebar that gives concrete much of its strength. As the salt causes the steel rebar to rust, the rust expands powerfully, sometimes blowing the concrete apart in a process called "rust-jacking." Stone bridges, particularly those built of granite, are immune to such problems.

Although extremely rare, under certain conditions, even granite can fail from salt attack. The elaborately carved Egyptian obelisk known as Cleopatra's Needle, currently residing in New York's Central Park, offers a painful example. In Egypt's mild climate, the coarse-grained syenite granite had weathered at a rate no more than three-quarters of an inch in a thousand years. But within just three years, installed in a harsh freeze-thaw climate, the granite began to exfoliate. Within eighty years, more damage was done than in the last two thousand. In the hundreds of years that the obelisk lay toppled in the Nile mud, it absorbed salts into its open grain structure. Those salts grew and expanded in the needle's new location, effectively erasing the deeply "inscribed names and mighty deeds of Queen Hatshepsut, Thutmose III, and Ramses II" carved at its base.[46]

Delaminating Dangers

Continuing to think about our first Rule of Bondwork, which specifies grain orientation, it is now clear why the grain of stone must be positioned perpendicular to forces at work in a wall. In the stone's horizontal alignment, the correctly fabricated and placed block maximizes the inherent strength of the material. Visually, of course, we know this. Seeing stone applied vertically seems uncomfortable, as if a paving pattern somehow climbed up a wall. Following the highest and best use of the material means truly bedding the stone, literally tucking it in, making it comfortable (stable) and putting it "to bed" to mimic the original orientation at the time of its creation. By aligning the material perpendicular to gravity, we return it to its primary relationship with the Earth: the very best position to resist forces pressing upon it.[47]

When the first Rule of Bondwork is ignored, the result is called "face bedding." Once almost unheard of, today the problem of face bedding is even more widespread than when unskilled labor quarried brownstone for the Northeast market. With few real quarries operating in the United States, most of the building stone sold today is collected with front-end loaders and pry bars rather than quarried with proper grain orientation according to the Rules of Bondwork. Typically collected in three- to five-inch thicknesses, the stone may be mortared against walls of dimensional lumber sheathed with plywood, its application sometimes ending inexplicably at corners or arbitrary divisions. Such collected materials—as opposed to quarried materials—are typically installed "flipped on edge," face-bedded to maximize square footage. This, of course, places the grain in the wrong orientation, setting the stage for future failure, all in a money-saving attempt to stretch the least amount of stone around what is called the fabric of the building.[48]

Although face bedding is aided and abetted by ignorance and poor training, the true culprit is economic expediency. It is much less effort (read: cost) to saw the stone across the grain and then use the primary grain to provide the split natural texture than it is to split the stone perpendicularly to the bedding plane and pitch off marks left by the wedges or saw. A case in point are the massive two-story walls ringing Empire State Plaza in Albany, New York, finished in 1976. Constructed of bluestone, a hard, fine-grained quartz sandstone native to the state, the entire project was face-bedded—the grain oriented vertically rather than horizontally—and is now in catastrophic failure.

Whole sections of stone wall are spalling off into the pedestrian walkway, putting citizens at mortal risk. These crumbling walls, which should have provided centuries of beauty and safety, have now been wrapped in chain-link fencing, putting off the problem of replacement to future generations.

Working with collected, rather than quarried, material, we are often unable to determine the stone's grain orientation, which we now know is required to build safe and strong stone structures. Equally important, collected material rarely can provide blocks large enough to furnish the required copings, lintels, and sills. Each of these architectural elements plays a part in keeping water from undermining a structure.

Follow the Building Forces

Of almost equal importance is how these pieces support the rich visual detail, the punctuation and grammar of architectural expression. Alain de Botton reminds us that just as "the alteration of a single word can change the whole sense of a poem, so, too, can our impression of a house be transformed when a straight limestone lintel is exchanged for a fractionally curved brick one."[49] When a window or building opening is spanned by neither a single stone (a lintel) nor a fractionally curved brick one but rather a collection of irregular, nonquarried stones floating inexplicably over the void, the very basis of our visual reference becomes dulled.[50] It is strangely unsettling to see stone appearing to float above an opening—and one sure hallmark of unconsidered masonry.[51] In the vernacular masonry of most cultures, when stone could not provide a large enough spanning member, the more complex geometry of an arch was regularly employed. Even a wooden spanning member, doomed to disintegrate and fail over time, is more visually satisfying than the current formula of floating stone facing. Missing a single stone detail such as a spanning window lintel irrevocably compromises the essential visual safety promised by the material itself.

When the skills to quarry longer-spanning members had yet to develop, ancient builders turned to wood to bridge building openings. Such wood lintels would often succumb to insect damage or rot and required frequent replacement. Masvingo, Zimbabwe

Whenever the forces of the building change direction, the first Rule of Bondwork reminds us that we must follow these forces and adapt our fabrication and setting. For example, a true arch converts the thrust around the opening of a window or door through the geometry of the wedge-shaped pieces, the voussoirs. Therefore, the grain of the stone must be kept perpendicular to the forces it will receive to maximize the stone's strength. The primary grain of the stone will therefore follow the forces within the arch, slowly changing axis as the forces shift.[52] Further, the

TOP
The skills required to quarry and set stone lintels to span building penetrations such as doorways and windows mark a significant milestone in an emerging culture. Giza, Egypt

BOTTOM
When spanning lintels are not available due to quarry or material limitations, masons may turn to a variety of other strategies, including this “jack arch” or “flat arch” in granite. Cape Town, South Africa

A stunning improvement to the "jack arch" lintel employs the "joggled" or "rebated" voussoir. Invented by the Romans, the joggle on the downward line of the voussoir stone strengthens the assembly for earthquake-prone environments. Jerusalem, Israel

OPPOSITE
One of the first (third century BCE) and best-preserved true stone arches marks the entrance to the fortification at Falerii Novi, south of Rome. Called Jupiter's Gate for the delicate carving that adorns the keystone, it advances one of masonry's essential forms that has remained in common use for more than two thousand years.
Falerii Novi, Italy

TOP
When the interior radius (intrados) of an arch follows a different curve than the exterior (extrados), the arch is called Florentine, after the city that perfected it.
Montepulciano, Italy

BOTTOM
Giovinazzo, Italy

joints of the voussoirs must remain perpendicular to the same forces. Rowland J. Mainstone writes that "ideally, the joints should all be at right angles to the compression, though there will always be enough friction to resist some obliquity of load."[53]

Arches work because the wedge-shaped voussoirs maximize the frictional area. In fact, the depth and height of voussoirs greatly increase the strength of the arch.[54] To intensify the friction, joints are kept to an absolute minimum, and care must be taken in their fabrication. It is no wonder, then, that while appearing in Egypt as early as the Twenty-Sixth Dynasty (1782–1550 BCE), the proliferation and ultimate perfection of the arch as a form did not arrive until the hardy iron tools of the Romans were more widely available. The evolving shape of the arch form can also be traced by the path of tool evolution. The typical semicircular arch perfected by the Romans repeats the same shape and geometry through every voussoir and is therefore simpler to fabricate. Alternatively, the gradually pointed arch of the Gothic period requires slightly differing angles for each location on the curve. This means that each stone on the Gothic arch is a unique shape. While it may have been technically possible to carve and repeat these forms with exactitude using hammerstones and early bronze chisels, the relatively few examples that do exist speak to the extreme difficulty of doing so.

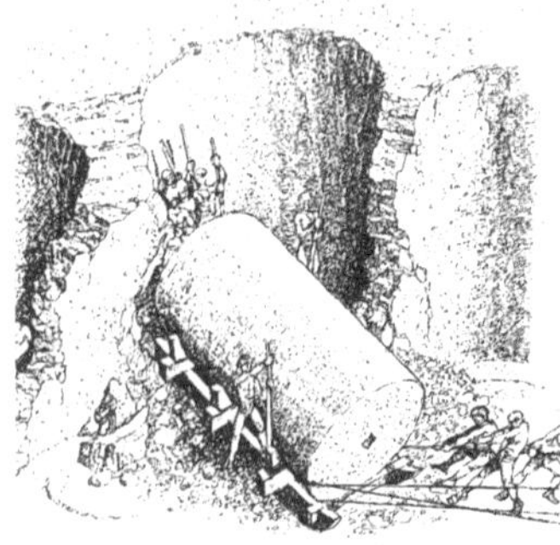

Even today with modern technology and methods, to quarry a column with correct grain orientation is no simple task. In fact, few if any modern stone manufacturers attempt it. Despite this somewhat uncomfortable modern truth, the ancient Cusa quarries in Sicily provide ample evidence that bedding the stone properly within the column was critical to structural support. To put this in perspective, the ancient column drums at Selinus, Italy, measure more than 11 feet in diameter, extend 53 feet tall, and supported monolithic architraves weighing more than 77 tons each.

Thinking about grain direction in the fabrication of stone columns likewise provides insight into how we weigh and trade design details against the practical considerations of what nature provides. To keep the forces perpendicular to the primary grain, a stone column must be quarried vertically so that the primary grain runs perpendicular to the column shaft. Although there are several examples of correctly quarried columns, the extensive labor and unusual geological formation required to procure living rock in such uninterrupted lengths led to the development of the segmented column.

The difficulty of breaking out a block vertically oriented (as a column requires) from the living rock is a task made more difficult by higher strength of the secondary and tertiary grain faces (also called "end grain") securing the block to the earth. In this context, it becomes easy to understand why many builders choose to risk the chance of column delamination rather than face the exceptional difficulty of quarrying vertical blocks. The historical record provides many examples of one-piece columns quarried and installed with vertical rather than horizontal grain orientation. The relative strength of the grain provides the answer. If the primary and secondary grain strengths are similar—as in high-density limestones and marbles—column shafts of single lengths with vertical grain may suffice. That said, getting it wrong can have catastrophic results. The modern restoration of Whig Hall at Princeton University in 2004 was initiated because bases and plinths supporting the six megalithic columns of the facade were failing in situ. Subsequent testing revealed that these bases and plinths had been quarried with the wrong grain orientation. Further, while each of the single-piece columns weighed approximately 14 tons, the load from the pediment exceeded 15.5 tons, meaning that the combined load on these bases was approximately 169 tons.[55] As US Senator Everett Dirksen once quipped about the national budget: "A billion here, a billion there, pretty soon it adds up to real money." The same is true of the loads on stone.

Single and segmented shafts in marble columns tell a similar story, since marble's primary and secondary grains are significantly closer in strength than in other kinds of stone. As previously mentioned, limestone's layers have been crystallized as they were subjected to intense pressure and heat during the mountain-building events that followed their sedimentation. The subsequent metamorphic process evened out the relative grain strengths and made the limestone hard enough to polish by hand, when rubbing with tin oxide. This is the source of the old definition of marble as "limestone that could hold a polish."[56] The geological metamorphic process also created enough compressive strength for single-shaft columns in support of medium loads. This having been said, contiguous marble blocks of larger size are rare, being more difficult to quarry and unlikely to be found in nature. Thus, marble columns supporting any significant weight or building forces are almost always segmented, their primary grain running horizontal to the Earth, honoring the dictum, "Lay all stones in their natural bed."

Even when columns of great length are successfully quarried, the difficulty of fabrication, transporting them to the site, and raising them vertically can be fraught. Eight granite shafts intended for the unfinished Saint John the Divine cathedral in New York were originally quarried and hand shaped, their final polishing accomplished on a giant stone lathe in Vinalhaven, Maine, in 1903. Although the design called for monolithic shafts of granite six feet in diameter and fifty-four feet in length, no one in the United States could manage it (although granite shafts measuring almost seven feet in diameter and sixty-five feet tall

were produced for Saint Isaac's in St. Petersburg, Russia). At Vinalhaven, three of the single shafts broke during fabrication.[57] The compromise was to quarry and polish the columns as two separate pieces (duoliths), one weighing ninety tons capped by a smaller segment of forty tons. These arrived by barge at the 134th Street dock, where they were hoisted onto solid-wheeled carts and winched by steam power to the cathedral site. Their safe erection remains one of the outstanding achievements of the structure.[58]

All of these examples demonstrate the importance of knowing the full extent of building requirements prior to choosing building stone. Knowledge that the grain must be aligned horizontally or vertically due to dictates of structure suggests selecting stone where primary and secondary grain strengths are closely matched. This includes high-density limestone that through extreme heat or pressure (or both) had begun to metamorphose geologically into greater density, and therefore greater strength. For the same reason, building with granite offers a safe way to mitigate the inherent weakness of the different grain structures. While there is a significant difference between the primary and secondary grain structures of granite, their relative strength is so high that in practice neither face will delaminate or fail.

Egyptian obelisks offer an interesting case in point. Their survival over the millennia is due to the incredible strength bonding the three grain structures. Exposed to weather with the primary grain oriented vertically, obelisks stand like a deck of cards. To survive for the ages, the primary grain structure must be strong enough to resist delamination in heat and cold, but still weak enough to allow it to be separated from the larger lithic mass in the quarry.

At the quarry site in Aswan, Egypt, we can still see evidence of the methods used to produce the obelisks. After using hammerstones of diorite to pound out long horizontal troughs to free the stone around its perimeter, one final step remained: the separation of the shaft along its length. This final area of contact was, by design, along the primary grain, and therefore the weakest connection to earth. Although the amount of time that was required to expose the four sides was likely measured in seasons, not months, I cannot imagine that the quarry master ever felt totally prepared to begin the final step of freeing the bottom of the stone across its primary grain or rift. Yet proceed they did, likely using simple wedges of wood that would slowly swell when

OPPOSITE
The 2,400-year weathering of the limestone reveals the horizontal grain of these unfinished segmented columns. Segesta, Italy

OPPOSITE, CLOCKWISE FROM TOP LEFT
When columns carry significant weight, they are typically segmented so as to keep the primary grain orientation horizontal to the Earth. In this way, their primary strength remains perpendicular to the forces traveling through the stone assembly. These striated column drums clearly show the grain of the stone perpendicular to the compression forces. Istanbul, Turkey

Segmented Doric marble columns with decorative fluting support the dome of the District of Columbia War Memorial. The fluting of the columns is typically carved in situ, since protecting the many edges (called "arrises") during transport and installation would be difficult, if not impossible. Washington, DC

This column, with its grain running vertically up the column shaft, would be considerably weaker in compression and may offer some insight as to why it has been taken out of service and relegated to garden ornament. Istanbul, Turkey

These segmented limestone columns still show the projecting bosses left from their initial placement almost two thousand years ago. Once the assembly was completed, the bosses would ordinarily be carved down flush with the finished surface. The stone beams spanning between columns are called architraves. Jerash, Jordan

the surrounding trough was flooded with water. As the wedges expanded, they created enough pressure to crack and separate the shaft from the living rock.[59] The key to this bold effort was ultimately dependent on the exact alignment of the primary grain with the forces of separation.

The only monolith of modern times that comes close to rivaling the achievement of the Egyptians was crafted for the 1893 World's Columbian Exposition of Chicago by the Prentice Brownstone Quarries in Wisconsin. Measuring 115 feet in length, ten feet square at the base, and four feet square at the top, it was separated from the primary grain of the quarry in exactly the same method as the obelisks of the Aswan quarry. Five "steam channelers" toiled away around the perimeter for three and a half months, "all but the loosening of the bottom of the stone from its bed. This was accomplished by steel wedges."[60]

Bedding the Stone and the Rise of Pattern

Form becomes effective simply as the result of structural combinations.
—EUGÈNE EMMANUEL VIOLLET-LE-DUC[61]

As we have seen, bedding the stone with the grain horizontal to the Earth is critical at the piece level to take advantage of the material's latent strength and set it perpendicular to the forces of the building. But there is another critical aspect of this elemental Rule of Bondwork that leads us into the fascinating world of patternmaking, one of the most soul-satisfying aspects of stone-masonry. Essentially, "bedding the stone" also requires that the ultimate shape of individual pieces be set horizontally in a longer—rather than taller—aspect. The basis of this rule is practical. Horizontal stones are less likely to tip over than stones set vertically. Vertical stones have smaller bases and are therefore significantly less stable. Furthermore, since a proper stone wall does not rely on mortar for strength or stability, bedding each stone works to maximize the inherent stability of each piece as we build upward. We intuitively grasp the stability of the horizontal. This is why ashlar patterns with rectangular shapes oriented vertically invariably read as pavements, as if someone had tipped the ground plane up onto the wall surface, giving it a disorienting perspective—a common fault of contemporary masonry practice.

Verticality remains central to stone architectural design. Pointing architecture toward the heavens has always been one

The Inca stoneworkers used projecting bosses extensively for handling the massive blocks that routinely weighed in excess of fifty tons. Their continued presence is testament to the unfinished nature of the work, interrupted by the arrival of the Spanish. Usually, the bosses would be dressed down and removed once the structure was completed. Ollantaytambo, Peru

One of the earliest examples of projecting bosses can be found on the unfinished granite casing stones that once completely covered the Pyramid of Menkaure at Giza. Luxor, Egypt

TOP

The attraction of the vertical form is unmistakable. The Rules of Bondwork do not eschew the vertical; they insist only that the vertical be attained through the assembly of stone fundamentally bedded in the horizontal aspect. Merzouga, Morocco

BOTTOM

Holding wrought coursework level not only keeps the bedding of the stone aligned with its highest strength but also works to key the wall into the steeply sloping terrain. Huairou, China

of its universal goals. Significantly, six of the Seven Wonders of the Ancient World were made of stone, and all of them pushed the upward limits of verticality to create awe and wonder.[62] Nothing was thought magnificent that was not "high beyond measure," as Sir Christopher Wren described it, and to this day, most architects default toward the vertical.[63] In the twentieth century, perhaps only the US architect Frank Lloyd Wright understood the equal power of the horizontal. The fact of the matter is, however, that the Rules of Bondwork do not eschew the vertical; they insist only that the vertical be attained through the assembly of stone fundamentally bedded in the horizontal. We should remember here that despite the collapse of the nave of the Beauvais Cathedral—the largest cathedral ever attempted—the vertical limit of horizontally organized stone has never been ascertained. The Rule of bedding the stone, then, simply helps to remind us that the stone fabric of a building must retain a horizontal character.

Extending the concept of bedding the stone, we must also bed our stone structures with the same intention as we bed the individual stones. We see this most clearly demonstrated when buildings or walls cut through changing grades. As the ground rises or falls with the natural landscape, gravity is pulling the wall or structure toward the earth, no matter how the grades supporting the structure may change.

The same rule applies when the slope is human-made. For example, a coping on a sloped wall or sloped roof gable must be

LEFT
Steep gables rigorously engage with the larger pattern to secure purchase. One can never rely solely on the bond of the mortar to hold the stone where life safety is concerned. Nuremberg, Germany

RIGHT
Borobudur, Indonesia

firmly horizontally bedded so as not to slide off, since, over the longer term, it is not enough to rely on the strength of the mortar to hold it in position. Essentially, the slope of the wall decreases the friction of the stone coping's dead weight, putting it at risk of slipping. By incorporating the principle of bedding the stone, the forces of gravity remain fully engaged to pull the stone downward, effectively locking it in place. Bedding the stone on its horizontal axis is a relatively simple strategy, yet it fulfills the twin goals of building safely and unlocking core attributes of stone as a building material. Successfully integrated, these strategies ensure that stone architecture will last not decades but millennia.

For more than fifteen years I was involved in salvaging architectural building stone from demolition projects in cultures racing to upgrade their built environments. Taking down these ancient walls introduced me firsthand to the many methods of structural bonding. In practical terms, this meant that each stone was carefully removed in the reverse order from which it was originally set, perhaps like reading a book backward. In taking these walls apart, I discovered many instances of what appeared to be a regular scattering of stones set in a vertical orientation. Such vertical stones would seem to contradict the critical horizontal bedding of stone on the piece level.

But I soon understood that the vertical pieces that seemed to contradict this primary Rule of Bondwork were simply part of a technique to tie the wall to the hillside. Vertically set pieces were

LEFT
The fully bedded caps at the Saqqara complex incorporate an overlapping key that may have inhibited their theft over the millennia. Apparently, this bedding technique is one of the oldest in the Rules of Bondwork, since Saqqara is considered the first work of stone architecture. Saqqara, Egypt

RIGHT
Cape Town, South Africa

This irregular ashlar pattern (tightly fitted and coursed stonework of irregular sizing) from the twelfth century is calming and strong. The structural and visual security is derived from full frictional strength supporting the fully bedded, horizontal aspect of the rectangular shapes. Jaisalmer, India

TOP
The proper horizontal bedding of the stone imparts a sense of stability and strength, despite the limitation of the local limestone's small piece size. Coustellet, France

BOTTOM
Walls of great height may be achieved by stacking and ordering stones horizontally. When each stone is properly "bedded," we say it is "at rest," achieving a position of stability capable of enduring successive civilizations. Machu Picchu, Peru

TOP
Often, when we find stone set in a vertical aspect, we are mistaking the block's end for its face. In this photo, the stones that show vertical orientation are working to tie the wall into the hillside. These are known as "lacing courses" because they stitch the wall and the retained hillside together, giving great strength to the assembly. Wanxian, China

BOTTOM
Lacing courses are composed of stones called "dead-heads" or "tie stones" that extend back from the face of the wall into the retained grade. Typically, such stones extend three to five feet into the earth they are retaining. Yangtze River, China

ABOVE AND OPPOSITE
Two visits spaced four months apart to Jinan in 2008 revealed the shocking theft of the fine ashlar blocks of a rural temple. All that remained upon the second visit were the tie stones and the structural hearting (the loose stone packed at the center of the wall). Jinan, China

almost always the butt ends of much larger stones returning into the retained earth. Known as "dead-heads" or "tie stones," such larger pieces are, in fact, properly bedded. By turning their longest face perpendicular to the wall surface, they bind the facing of the wall to whatever is being retained. It is not uncommon to see these tie stones running in series called "lacing" or "bonding" courses. In structural brickwork, this lacing occurs every sixth course and is designed to weave or bond the brick courses (or wythes) together.[64]

Earliest Experiments

The line in the sand, so to speak, of where stone architecture began is difficult to draw and is arbitrary. Certainly, early humankind pulled stones into circles to create sacred or ceremonial spaces, celebrating what then happened in the middle. We moved from delineated spaces to buildings by adding walls and roofs, from buildings to temples with their inner sanctums and hallowed volumes. The settling of nomadic peoples and the gradual turn to agriculture prompted this awakening. Turning over the soil produced stockpiles of loose stones, aggregated at the field's edges. Stacked into primitive walls, they enclosed livestock and formed the basis for the larger concept of boundary—those within and those outside. Agrarian prosperity fostered larger populations and allowed for infrastructural investments. The US sociologist Joseph Henrich points out that as groups "began investing in certain crops in fertile regions, people needed to be able to secure and hold land.... This gave a substantial edge to groups with any social norms, including rituals or religious beliefs, that made them better able to defend territory."[65] Ceremony and ritual developed to reinforce the collective, creating a spiritual premise and mythos to reinforce the vital concept of the community staying together. William Richard Lethaby cites these ceremonial commitments and credits them with the rise of monolithism and the eventual worship of deities through stone.[66]

Recent discoveries at the sophisticated temples at Göbekli Tepe support this hypothesis. Located in the far eastern corner of Turkey, more than two hundred T-shaped, monolithic, rectangular columns were erected and mounted into depressions carved into the living rock. The columns form the skeleton of an architectural framework creating roughly circular or oval spaces. Many of the columns have decorative animal-figure carvings. Although preliminary, the dating of Göbekli Tepe appears to be as early as

ABOVE
The mortuary complex at Saqqara, thought to be the earliest example of wrought stonework, has survived 4,700 years of tumult on the alluvial plain of ancient Egypt. Note how the integrated cornerstones (single stones between wall and expressed column) weave the differing planes together. Saqqara, Egypt

the tenth century BCE. Such discoveries challenge our long-held chronologies of where and when architecture began.

The columns of Göbekli Tepe aside, most of the earliest remaining wrought stonework (and the Rules of Bondwork themselves) owes its origin to the temples of ancient Egypt. So, it is of great interest to see details worked out in these structures and follow their experimentation in both form and pattern. By studying the Nile civilization's emerging forms—diverse as the mud-brick catenary arch and the post-and-lintel roof—we glimpse how comfortable the stone design mavericks of ancient Egypt were with their emerging techniques. They provide a road map for how we might negotiate our present dilemma in which technical achievement of new forms remains thinly connected to stone's highest and best function.[67]

Our knowledge of the complexities of pharaonic society is, of course, imperfect. Meanwhile, its accomplishments in stone tower above us, and to this day remain largely unexplained. The stone facade at Saqqara is difficult to reconcile with the mud-brick mastabas that preceded it or were built at the same time. At Saqqara, the limestone blocks rest in full frictional contact; their integral cornerstones weave the planes together, forming

The post-and-lintel technology of most Egyptian architecture limited interior spaces to the small volumes available between the supporting columns. The free span of an unsupported stone lintel is extremely limited. Extending the columns vertically helped create an illusion of space, but the lived experience, until the later invention of the vault and dome, was more like a dense forest of massive tree trunks. Luxor, Egypt a cohesive mass that has endured the ravages of succeeding civilizations virtually intact. The shadow play of the simple relief charges the form with visual interest, ever shifting as the sun traverses the open sky. The structure follows—seemingly invents and codifies—all of the Rules of Bondwork. So where are the failed experiments? Surely some halting stacking of stone must have occurred before this temple arose from the sands of the Sahara. Saqqara reveals a gap in its masonry record, a gap between mud brick of the past and the exquisite wrought limestone that followed, unannounced and unheralded. Lethaby reminds us that "to a large degree, architecture of wrought stone is an Egyptian art."[68] Perhaps Saqqara stands alone because earlier masonry failures would naturally vanish, their collapsing stones reused and absorbed in the eternal life cycle of birth, death, and rebirth.

Ironically, it was this preoccupation with death and rebirth that provided the impetus to turn to stone in the first place. Saqqara was built, after all, as a mortuary complex to honor and ritualize the king and his descendants passing on to the afterlife. The building program called for timeless materials and craftsmanship capable of speaking to all future generations. In the arid desert climate, where the rain's natural acidity would never break down the high-density limestone of its walls, the material must be fairly viewed as possessing ultimate strength. Saqqara mates the strongest material with ultimate workmanship, creating a transcendent unity. The results demonstrate the equal partnership between materials and the Rules of Bondwork, a partnership that produced a fine and cohesive whole. Neither the material, nor the form, nor the craft draws undue attention to itself; rather, they work in balance to express harmony and represent the first enduring work of great art in architecture. It is useful to remember that Egyptian society evolved over a period of almost nine thousand years, and that the towering achievement of Saqqara occurred about halfway through that evolution. Demonstrably guided by what would become the Rules of Bondwork, Saqqara's carefully wrought stone walls demonstrate a massive sophistication and relay a central truth: aesthetic triumph arises directly from innate structural success.

Unsurprisingly, this achievement was not universally understood or accepted, as illustrated by the original walls buttressing the burial temple of Hatshepsut at ancient Thebes in the West Nile valley. Built some nine hundred years after the success of Saqqara, the pattern is coursed and tightly wrought, but the individual pieces are set vertically rather than in their natural bed, as the nascent Rules of Bondwork would insist. Only four to five courses of this example remain—those that were protected under the shifting sands of the Sahara. The rest were pried loose and have long since been carted away to build other structures. The inherent weakness of the vertical pattern may well explain why so few courses of this ruin remain. But the more interesting question to ask is whether the verticality of this work, which contradicted the first of the Rules of Bondwork, was built out of ignorance or with intent. That is, does this deviation represent ignorance of the Rules, or does it provide one of the first examples of bending the rules for structural success against the expressive concern of the larger complex? Was the stone intentionally designed to be oriented heavenward, standing at attention, sentinel to the tomb and the contents it was built to protect and honor?

The 3,600-year-old wall forming the battered retainage to the Hatshepsut's West Nile mortuary temple at ancient Thebes has an almost modern character, since the coursework is vertical rather than horizontal. It was assembled during a time when the Rules of Bondwork were first being codified by the ancient Egyptians. Unfortunately, only five courses of this magnificent wall (the lower section) have survived. Luxor, Egypt

Beyond the Orthogonal

Bedding stone horizontally becomes even more critical with oval or polygonal shapes, where the geometry of the piece can diminish the stability of the overall pattern. Nowhere is this more obvious than in river rock work. Stacking rounded shapes is a relatively new phenomenon made possible by increased mortar strengths. The modern method, rightly dismissed as "stacking marbles," relies on the bonding of the rounded surface to the substrate. At best, it is an overused visual reference to the early settlers of the western United States who would take building stone from the local creeks in their rush for protection before the coming winter. But even pioneers had the structural sense to choose misshapen river stones rather than the smoothest and most rounded. In the best instances of this pattern, we find a mix of round river stones and more trapezoidal materials, which brings increased stability to the assembly. A successful early twentieth-century example of mixing round river rocks with more rectangular shapes common to brickwork is found in the masonry of Charles and Henry Greene, brother architects working extensively in Southern California at the turn of the last century.

Some of the very best work employing river rock splits the river stones in half to reveal their inner beauty but sets the material horizontally, in observance of the Rules. The stability of the pattern is supported and bedded as safely as possible, while the expressive potential of the material is fully realized.

TOP
The architectural team of Greene and Greene successfully combined river rock with brickwork (above the roofline in this case). The Gamble House disguises this brickwork under a structural render or plaster that unifies the color of the assembly. Note the horizontal aspect of the river stones. Pasadena, California

BOTTOM
Fluvial stones, whose rounded forms were created by interaction with water or ice, are commonly called river stones. Here they have been split open to create a flat wall plane. Note the rigorous horizontal aspect of the worked stone, as the bedding of the shape, if not the internal grain, is honored. Atlas Mountains, Morocco

While work created from river stone can be exceptionally beautiful, the potential damage to the environment must be registered. Although deposits of fluvial stone can be found on land in many areas, sometimes river stones are collected from the stream itself. On almost every continent, I have found streams and beaches denuded of habitat by the removal of their rounded river stone or flagstone. River stone is a finite commodity whose end is rapidly approaching. In 1990, returning to one of my favorite streams south of Siena, I found it irrevocably altered. This same stream had provided countless evenings of bathing pleasure as I splashed the mason's dust and sweat from my weary body. But in the ten years since my apprenticeship in the early 1980s, small front-end loaders had driven up the streambed, prying loose and forcibly removing every stone of commercial value. This process widened and quickened the waterway and reminded me of the many creeks destroyed in the western United States by overgrazing cattle, a tragedy being repeated on private lands in Idaho, Colorado, and Montana today.[69]

Stripping the stone from a steam for commercial resale is very different from collecting surface boulders to split and build with. Humankind has cleared land of rock and boulders to make it usable since we first traded a hunting spear for a phallic dibble (the tool for inserting seeds into the womb of the earth). This collecting of surface stones, called "grazing," is still a viable way of gathering material for projects. In fact, many times in the western United States you will find that landowners have pre-emptively cleared the boulders into large piles. In my experience, after politely asking permission, I have been encouraged to take "as many as possible!" Of course, such collections are "dead stone," since they have been excavated and dried in piles, and are no longer coursing with groundwater. Although I often split such boulders by hand, for reasons previously explained, dead stone is more typically processed with mechanical diamond saws rather than carved or shaped with human labor.

An important subset of the first Rule, "lay all stones in their natural bed," continues with the caution, "never on their points." If bedding the stone in its horizontal aspect maximizes the visual and structural stability of the assembly, continuing to bed this material horizontally when shapes are polygonal or irregular leverages the strength of the material. Even when working with irregular shapes, proper bedding is core to secure and stable stonework. "Never on their points" means that an irregular shape or polygon must never be oriented with the sharpest (most acute)

Proper bedding gives a stone wall stability, and in polygonal Incan stonework, the rule is strictly observed. Irregular shapes are seated flat on the bottom. This powerful bedding technique has allowed these walls to survive intact in one of the most active seismic environments on Earth. Nearly sixty earthquakes over 6.0 on the Richter scale have shaken these walls since their assembly in the early sixteenth century. Cuzco, Peru angle wedged downward. If the point should suddenly slip out of its position, creating a void, the entire wall might fail.

In any case, the best craftspeople shun the fragile geometry of sharply pointed stones, avoiding pattern voids with careful piece selection and shaping. As it turns out, one of the central skills of the mason's craft manifests itself in polygonal pattern-making. Like a photographer who compositionally frames the negative space, a stonemason anticipates the shape of the coming piece while crafting the current one. In this way, masons not only shape the bottom and side that will abut the work already in place, but also craft the angles of the sides still free so as to accept and bed the pieces that will follow. In the best work, this preparation for the next piece shape avoids angles of less than 52 degrees to ensure that the shape of the following stone will not be too sharp or vulnerable.[70] Anticipating the coming shape and bedding the stone to avoid the sharp angles that would lead to "setting on their points" is central to the journeyman mason's skill set.

The most skilled masons reduce the amount of site-fabricated labor by intuitively reaching for the stone shape best suited for the position. Sounds simple, doesn't it? What might be actually happening is much more complex. During my own apprenticeship, my teachers in the guild spoke of physically relaxing into the knowledge of the shape while laying polygonal stonework. They intimated that our unconscious mind, working with our innate visual memory, has already cataloged the shapes of stones available in the work area. By relaxing into that certainty, the separation between what is unconscious and what is conscious grows thinner, and one reaches for the best stone quickly and intuitively. Although I never mastered this intuitive ease during my apprenticeship in Italy, a decade later, I frequently—if somewhat inconsistently—recognized what my teachers had described. I most commonly accessed the unconscious in a state of fatigue and quiet, finding unexpected fluidity late in the day. Those moments when I achieved that expansive mind state marked some of my peak experiences in working with the tools. Because such a state is achieved only when my surroundings are relatively still, I rarely allow radios or music on my jobsites. Setting up the conditions favorable to achieving mastery turns out to be an important step toward achieving it. Although there can be no certainty of success, preparation and quiet maximize the possibility of recognizing and inhabiting the experience, should it come.

LEFT
In Japan, stonework requiring the most strength will invariably be the most tightly wrought, illustrating the relationship between the architectural program and the size, scale, and even pattern of the assembled wall. All tightly fit stonework may be evaluated on a continuum of strength, with the tightest and most fitted work producing the strongest results. Constructed in a pattern of random rubble built in courses, this wall above would likely occupy the middle of the continuum, while the tight-fitting irregular ashlar (in the next image) represents the pinnacle of strength. Kyoto, Japan

RIGHT
Irregular coursed ashlar can produce walls of profound strength, since the finished work mates the friction of block to block connection with irregular locking geometry. Kyoto, Japan

I have found that this altered state can be taught and learned. The Buddhist monks of ancient Tibet, in fact, are reported to have used the practice of polygonal stonework as a teaching tool in the religious training of novitiates. Similarly, in the Japanese Shinto tradition, the state of flowing, centered connectedness allows access to the unconscious mind, where practitioners can connect to the "gestalt of the stone." In this way, the boundary between the human and the material dissolves in the wash of creative energy connecting all things.[71] As it turns out, releasing the mind for piece selection might turn out to be the same skill set required for being and nothingness.

In striving to lay stone courses in their proper bedding, sometimes the stone proves particularly intractable, or the available labor for careful crafting is limited for one reason or another. In such cases, small chinks of stone may be added to fill voids and facilitate the careful bedding of the following stone course. "Chinking," the process of adding smaller stones to shim or close the gap between two irregular pieces, relieves the need to work the stone shapes into perfect contact. While technically a shortcut, chinking is a legitimate technique to reduce the labor necessary to square or tightly fit irregular shapes. As long as the primary force within the wall remains downward and static, the wall remains stable. Chinked walls possess less strength, since they lack the full contact with surrounding stone that would normally serve to amplify the frictional force. But they still achieve strength of assembly through their mass and proper

Chinking walls with repurposed ceramic tile may have started as a practical way to build simple walls for agricultural workers unschooled in stone shaping and without basic tools. But their expressive results exceed many attempts by skilled masons using more rigorous methods. Shuitou, China bedding. In ancient Japanese fortifications, walls of lesser structural importance are often successfully chinked with visually stunning results. But when real programmatic strength is required, the chinking vanishes and the stonework is powerfully constructed in full contact and tightly wrought. This strategy is found in virtually all ancient cultures where the skills for such close-fitting work were developed.

Fully Wrought Dimensional Stone

In hard and durable building stone, tight joint work requires considerably more labor and is therefore more expensive. The cost of fitting irregular stones tightly together has always been considerable, even when the only payment was food to keep the workers alive. In the fifth century BCE, when Herodotus visited the Great Pyramid at Giza, he noted an inscription claiming that sixteen thousand talents of gold had been spent on "radishes, onions, and garlic" eaten by the one hundred thousand souls who labored on the structure for more than twenty years.[72] Lethaby cites this as an example of the "myth of cost." He implies that recording such an extraordinary cost by carving it into the monument itself amounted to an early example of ostentatious wealth, or what we might today call "billionaire bragging rights." In fact, from my perspective, that sixteen thousand talents of gold was surely a fraction of the construction cost.[73] And today, not only is it unacceptable to pay our workers in garlic and onions, but the permitting process alone would likely exceed the twenty-year construction schedule.

The Great Pyramid at Giza demonstrates the squaring off of stone blocks into regular courses. Once square, the blocks were assembled relatively quickly, resulting in a much-reduced building cost. It is estimated that during the twenty-year assembly of the Great Pyramid, the Egyptian master builders placed an average of eight hundred metric tons of material every day; each of those blocks required quarrying, shaping, and transporting to the site prior to placement. The blocks were squared and abutted to an average tolerance of only half a millimeter (one-fiftieth of an inch), a specification unmatched even by today's computer-assisted manufacturing.[74] It is an astounding achievement by any measure. To have attempted the work using an irregular coursing would have cost much more. Although an oversimplification, the regular courses allowed less skilled labor to manage the assembly work, since the stacking of regular objects is a relatively straightforward task. For this reason, brick rather than stone provides

The last of the surviving Seven Wonders of the Ancient World, the Great Pyramid at Giza remains an unparalleled feat of engineering, logistics, and workmanship. The tallest human-made structure on Earth for more than 3,800 years, it contains an estimated 2.3 million blocks. Giza, Egypt

Although the Egyptian builders turned to "battered" walls for their natural stability and strength, they are also visually comforting. In this example, the vertical joints often rise obliquely, stitching the blocks together within their respective courses, as a way to further strengthen the assembly. Giza, Egypt the usual gateway into the craft of masonry. Only after the basic skills are mastered with regular shapes can the assembly of dissimilar blocks and shapes be approached.

As mentioned in chapter 1, preparing regular coursed stonework in advance offers the added benefit of lowering the cost of transport. Since the material is shaped close to the quarry instead of near the site, the assembly of squared-off material eliminates transport of waste material that would result from shaping the blocks at the site. Simple economics, then, have worked to drive the patterning of much of the world's architecture. Working with squares and rectangles, we trade lower strength for simplicity of assembly and economy of freight. Because of this, the most common pattern of stonework is ashlar—different sizes of squares and rectangles blended, or in courses of running bond (regular or irregular rectangles assembled in rows). Random ashlar patterns (blends of different sizes) reduce material costs further by maximizing the yield from the quarry.

Although regular ashlar patterns are cost-effective and largely "strong enough," when walls must resist more extreme forces, other strategies have been developed. For example, the Egyptians discovered the effectiveness of breaking the verticality of the end joint of each block. These "oblique rising joints" are deployed when friction will be inadequate to prevent lateral displacement.[75] Egyptians also "battered," or leaned back, the wall itself. Pushed to its most extreme form, the battered wall becomes a pyramid, the most inherently stable form possible. The Irish architect and archaeologist Harold G. Leask describes how the batter evolved from its early Egyptian roots: "Though adopted for practical reasons in the first instance, the batter seems to have been retained in later works as a grace, an architectural refinement, giving not only a sense of stability but an aesthetic satisfaction."[76] To this I would only add that that same satisfaction is ripe for the space constraints of modern life. With the wall leaning back slightly, we are given more space to pass by and can't help but experience the urban world as a more generous place.

An architectural program's anticipated forces within a wall provide a primary determinant to the size of the joint, the pattern of the material, and the scale of individual pieces. Yet there are other strong reasons for tight joint workmanship. Consider for a moment the visual excitement created when stonework is tightly wrought. Much of the awe we experience in exceptional stone architecture traces its roots to the profound expression, the form or gestalt of the whole. From earliest times, stonework at a high level of craftsmanship speaks both to the power of coherence and to the coherence of power. The more intensely we craft stone—one of nature's core elements—we demonstrate our mastery not only over the stone but also over nature itself. From before recorded time, humanity has struggled against the incoherent, the seemingly random forces of nature. Stonework—read symbolically—converts every chisel stroke, every hammer blow, into an intention to survive. In some significant way, when we use stone to build a bridge, to ford a rushing river, or to build walls to protect against nature's or humankind's fury, we engage in a timeless commitment to survival.

Ruling elites have advantageously leveraged complex construction and labor-intensive logistics since the dawn of civilization. Keeping the populace fit and trim in peacetime reminds me a little of my own parenting strategy in raising my five daughters. An organized physical project or harder-than-anticipated bike ride with the family was often useful in directing the chaotic energy of my small tribe, and the resulting strong appetites and heavy slumber were an added boon. History offers countless examples of this "make-work" strategy, including the spectacular stonework added to our fledgling National Parks during the Great Depression.

One of my favorite examples comes from the Inca. To build a northern outpost at Villamarca on the road to Quito, the rulers sent some 450 finely cut ashlar blocks, each weighing between 450 and 1,540 pounds, more than a thousand miles. This wasn't a smooth path over a flat road, but rather a pitted walking track over the steep Andes. Of course, Villamarca had its own building stone of excellent quality, but one can only imagine the *sapa inca* (ruler) informing his master builders that only the stone from Cuzco would do. It is estimated that a work party of 4,500 would have taken half a year just to shift the stones.[77] In other examples, such as in czarist Russia, the impulse might have been less explicitly tied to making work for work's sake and more about striving for the (nearly) impossible as an expression of incomprehensible wealth and power.

To some degree, the same can be said for the awe inspired by the erection of ancient dolmens and menhirs weighing hundreds of tons. Whether we're speaking of relatively primitive tribes erecting single rough stones, or more refined civilizations raising a monument like a stele or obelisk, we admire earlier cultures marshaling their communities to achieve so singular

a goal. The epic aspect of ancient building achievements makes it is easy to understand monolithism, the worship of great stones.[78] The psychologist Marie-Louise von Franz concludes that "stone symbolizes what is perhaps the simplest and deepest experience—the experience of something eternal that man can have in those moments when he feels immortal and unalterable."[79] When we free a block of stone from the earth and move it into position, we have participated in one of the primary acts of civilization, and we have created another toehold on the planet for our children's children. When we give this freed block a "position of certainty," aligned with the successful strategies of the past, we have done our very best to ensure that our labor will survive for those same children to experience it.

Importantly, the first Rule of Bondwork—lay all stones in their natural bed—is foundational, both theoretically and practically. It acts as a gateway for other closely held secrets. In all cases, the Rules insist that if a structure is built mindfully, weighing the placement of pieces, crafting the fit, and finding the position of certainty for each individual component, the structure is likely to succeed. Such success supports the immediate use, long life, and longer-range programs of the civilization it becomes part of when finished. It also holds out the real possibility that the final assembly might transcend its own time, speaking forward to civilizations yet to come.

ABOVE
Most of the massive stones at Stonehenge were dragged more than three hundred miles before finding their "position of certainty" in about 3,000 BCE. The most interesting feature of the assembly, beyond its implications as a cosmic calendar, can be found at the tops of the upright members: the use of carved tenons for the capping lintels. Stonehenge, England

Chapter Four

WEAVING FOR STRENGTH

What is a structurally optimal form turns out to be a visually optimal form.

—EDWARD TUFTE[1]

Patterns result largely from the repetition of structural forms. They trace their reason for being to the mission of supporting and increasing the strength of the entire assembly. The pattern created by the stone material wrapping a structure is often referred to as the building fabric, and it is a fitting coincidence that written construction records of the late medieval period were written on fabric rolls recording the building accounts, materials used, and wages paid.[2] The fabric of a building, then, can be seen as linguistically demonstrating how the masons believed they were weaving their materials into a singular whole. In fact, "weave it together for strength" is the second of the Rules of Bondwork, and, like the first, its concern is specifically structural. Just as setting stone in its natural bed increases the strength of the assembly through the orientation of the material grain, weaving the material correctly into a cohesive fabric increases the bonded strength of the assembly by varying how the shapes are finally combined. In the natural world, we see the same principle at work when we find birds creating nests from dry grass; even without any adhesive agent, the form is strong enough to withstand the buffeting of winds and daily use. Indeed, weaving materials together is a successful strategy shared by both the human and the animal kingdoms, and numbers among a culture's basic skills.

It goes without saying that stone is not actually woven, since it is rigid. "Weaving stone" is a metaphor for the core practice of defining ideal stone shape and mandating how it becomes integrated into the mass of a building. The repetition of similar shapes provides patterns that determine the strength of the overall structure. "Bond" and "bondwork" are the practices of defining piece size and shape to which "the whole is bound into one compact mass."[3]

Within my mason's guild in Siena, we referred to the process of maximizing the strength of our structures through pattern as "bond strength," conceptually linking pattern to structure the same way we might expect to link two people in a relationship. Extending the metaphor, the stronger a couple's "bond strength," the more likely they are to withstand the vagaries of couplehood. The bond strength of a wall is not dissimilar. As an apprentice, I was always reminded that the strength of the bond that I created directly contributed to the strength of the building; every stone I set thus represented a decision affecting human safety. In fact, until masonry structures were replaced with stone veneers having only a tangential connection to the building's core, the skill of maximizing strength by weaving a bond was central to what it meant to be a stonemason. And because the way in which we weave stone together creates a pattern that directly supports a structure's ability to stand, patternmaking early on became central to the mason's craft.

Only after a decade of working with the tools did I come to understand how deeply the decisions of bond strength inform the larger pattern. For not only does bond strength support masonry safety, it also often acts as the driver of the pattern aesthetic. A simple example of the principle is the avoidance of small slivers of stone in our patterns. The issue is common when using pieces of fixed length. Visualize using bricks to lay a course from one fixed point to another. If we are to avoid slivers of brick, the ability to adjust the piece length (assuming we cannot shorten the wall) becomes key. Without this flexibility, the pattern must be interrupted by a sliver to accommodate the variation between the unit length and required dimension. In stone, where longer customized lengths can often be crafted, it is considered best practice to extend the length of the last piece, avoiding the unsightly sliver.[4] Slivers of stone are unattractive and introduce structural weakness into the larger assembly. Avoiding small slivers increases the strength of the assembly while protecting pattern continuity.

One way of looking at pattern is as "structure revealed," since the best patterns spring from real structural needs. Letting the structural concerns develop our pattern, we are turned away from the easy temptation of merely decorating. Instead, we are moved to build structural strength by weaving our stones together, wringing power from each block's interaction with the next. These woven patterns provide external skeletons to the structures we create, wrapping the building envelope—the building fabric—like a skin, safe for human habitation and regard. By focusing on the larger requirements of forming space, light, and structure with stone, we create pleasing patterns that are fundamentally honest, since our pattern develops directly in response to that larger, meta program. Pattern in true masonry practice is neither a graphic to be cut and pasted from our computer nor a rigorously applied geometry according to a set algorithm. Rather, pattern is the accumulation of individual decisions responding to structural demands.

An interesting case in point are the fine ancient buildings of Mycenae, Greece. These include the "beehive" tomb of Agamemnon (also known as the Treasury of Atreus), the Tomb of Clytemnestra, and the so-called Lion Gate. In each, the builders corbeled massive blocks over the openings to lessen the weight on giant spanning members, creating a triangular void where a lighter panel of bas-relief sculpture could be placed.[5] The resulting pattern, deriving directly from structural need, continues to capture the imagination thousands of years later.

Running Bond

Masonry succeeds as a building material in large part because of the key fundamental organizing principle of the natural world—gravity—which pulls stone downward in a frictive embrace. It would stand to reason, then, that the very strongest walls are those created using the very largest stones, stones that use their enormous weight to increase their frictive connection. As seen in the pyramids of Egypt, Cyclopean masonry is most typically assembled with the simplest patterns, because the oversize mass of the stones contributes so much strength to the larger structure. Bridging the inherent weakness where blocks abut is the simplest, yet greatest, structural need. Our instinct to overlap the joint of the stones below with a larger spanning member arises out of the need to address this weak point of interconnection and creates the first pattern of stonemasonry, known as the running bond. The running bond pattern succeeds

LEFT
The ashlar header blocks are set "headers and stretchers," with the former acting as bonders or "through stones," extending through the wall and tying the two faces together. Jinan, China

RIGHT
The main lintel spanning the entrance to the Tomb of Agamemnon has corbeled coursework above to lighten the lintel load. A bas-relief carving, long since lost, would have filled this void. Note the irregular coursed ashlar on the retaining walls of the approach. Mycenae, Greece

The third and smallest pyramid at Giza (only one-tenth the size of the Great Pyramid) was started by the pharaoh Menkaure and finished by Shepseskaf. In the twelfth century, a local sultan attempted to demolish it, but the tight fit of the stone blocks resisted their extraction. Instead, only the faces of the blocks were broken off, spoiling their shape. Giza, Egypt

The immense retaining wall supporting the Temple Mount in Jerusalem is constructed of a moderately hard limestone. The ashlar assembly derives much of its strength from the intense friction of the close-fitting configuration and the massive weight of the individual pieces. The chiseled borders that follow and finish the perimeter of each block face are known today as the "drafted margin." Jerusalem, Israel by consistently bridging and topping weakness with strength. The pyramids are "simply woven" by their running bond pattern. The simple rhythm of bridging weakness with strength, block after block, created a pattern that provided tremendous strength of assembly. Only radical forces could successfully dislodge its Cyclopean scale.

Running bond pattern is the most basic and, in many ways, the most effective way to weave regular-size stone blocks together. Since the standard shape of a stone block is roughly rectangular, rather than square, longer-than-tall stones may be seen as "running" along the wall in courses of similar height. When, as in the case of the pyramids, the bridging pattern of running bond is coupled with very tight joints, even more strength can be wrought. Crafting large blocks to fit so snugly formed the basis of Euripides's reference to a "red canon," of masons smearing adjacent, mating stone surfaces with red ocher chalk. When pushed together, the rubbed-off high spots indicated where further crafting was required. Signs of this technique can be found on much of the tightly wrought stonework in ancient Greece (Euripides's home) and even earlier in Egypt, where "on the stones may be seen the red spots of paint left from the testing by a reddened trail-plate as on the stone of Khufu, at Gizeh."[6] It is assumed that these "stones were worked flat and square by means of being tested against a plane surface smeared with paint."[7] In fact, the technique at the Giza pyramids—namely, hammerstones and simple copper tools—yielded the largest example of tight-joint masonry the world has ever known. The astonishing craftsmanship that these close joints demonstrate created intense friction and strength, and in four thousand years of abuse, destruction has been limited to the breaking off of block faces rather than their wholesale removal and theft. Friction-fit blocks are set so tightly in the pyramids that they cannot—or will not—be pried loose.

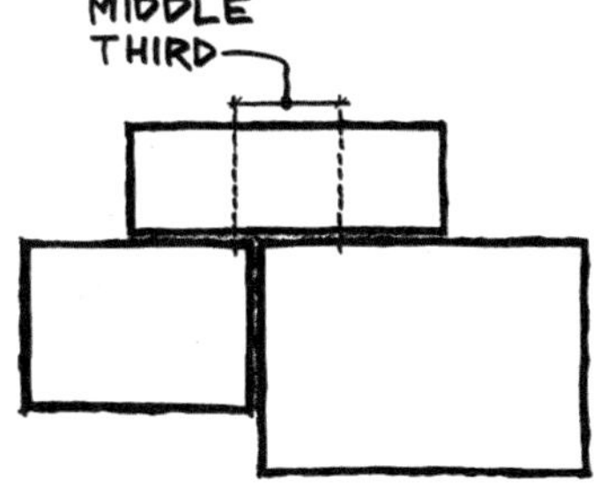

By making sure the joint being spanned falls into the middle third of the piece above, we have maintained the essential bond strength of the pattern while granting flexibility in the setting to maximize the material yield.

What Rowland J. Mainstone defined as "well lapped rising joints" the Freemasons guild called the "middle-third rule."[8] This rule lays out the acceptable range of joint flexibility that will not compromise the resulting bond. Most simply stated, the joint below must fall into the middle third of the stone being set above. This careful bonding allows the overlaying stone to be consistently woven over the breaking joint of the stone below and forms the baseline technique of weaving together for strength.

TOP
Apparently, the twelfth-century Khmer masons who built the temple complex at Angkor Wat did not fully comprehend the structural strength conveyed through woven patterns. When the one-third rule is not followed and each stone block is stacked directly on the stone below, the pattern is called "stacked bond" and is unstable. Angkor Wat, Cambodia

BOTTOM
Despite misalignment of the coursework, the rule of weaving the pattern for strength was strictly followed as this structure was extended. Havana, Cuba

Of Rubble and Ashlars

"Ashlar" and "rubble work" are the two most misused, most misunderstood terms among stone, design, and construction professionals. Essentially, tight-joint masonry (joints of one-eighth inch or less) of regular rectangles running in courses may be accurately described as ashlar work.[9] When the blocks have continuous joints, but the joints are not perfectly horizontal (or even vertical), the work is called irregular ashlar. Adding to the confusion, ashlars may also create random patterns of rectangles without continuous joints (random ashlar). What matters most in all of these variations is that, to be an ashlar, the shape must be essentially (if not precisely) rectangular and extremely close fitted or finely wrought.

When such rectangular blocks are not "finely dressed," the work is classified as rubble masonry. This difference between ashlar and rubble is a major dividing line in the pattern language of stone, and one worth taking the time to parse carefully. "Coursed rubble," sometimes called "random rubble built in courses," might look the same as "coursed ashlar," except for the wider joints. "Random rubble" is constructed of roughly rectangular shapes, loosely assembled (wider joints), and is not run in courses.

When the shapes of random rubble become too irregular, and the form moves into five-sided or even six-sided faces, the pattern description changes to "polygonal." Polygonal work is almost always tightly wrought, since, if one is going to the trouble of shaping the stone to fit the pattern, one might as well shape it to fit snugly. In this way, the strength of a polygonal configuration derives explicitly from the weaving of the pattern and from the superior powers of its cohesion. For this reason, it never superseded the strength of Greek fortifications where large blocks were trimmed to interlock.[10] Verbally describing the small differences between random rubble and polygonal stone patterns reminds me of the difficulty I have when I meet someone new and they ask me to describe my sculpture work. Some things are best revealed visually, and certainly pattern descriptions and their variations are better illustrated with photographs.

In broad terms, the wide category of rubble masonry—walling considered "not tightly wrought"—is less costly than ashlar, since the stones are less worked. Of course, the pattern is also significantly weaker than tight-fitting ashlar, since the friction from piece to piece is also reduced. True ashlar blocks are labor intensive and require great skill to maintain their maximum

Finely wrought ashlar blocks are the exception rather than the rule at Machu Picchu. Since much of the other work is irregular for strength, the regular coursing of the temple at the top of the ancient city creates an atmosphere of unexpected formality. Machu Picchu, Peru

The irregular coursed ashlar of the Inca occurs only in the most important sections of the royal compound. Not exactly squared, their faces were dressed and "pillowed" by hammerstones. Machu Picchu, Peru

The finely wrought ashlar masonry of this ruined edifice offers us important clues to its relative standing in an imperial Roman city. In virtually all cultures, the intense labor required to craft tightly fitted blocks communicated an exalted status. Jerash, Jordan one-eighth-inch joints between the stones. For the Inca, where all styles of stonework were tightly fitted, the ashlar pattern was reserved for royal use and important ceremonial buildings. The US anthropologist Dennis Ogburn adds: "This type of stonework played an important role in power relations between the state and its subjects.... They symbolized the immense amount of labor the Incas had at their command, which could be employed for any objective they chose."[11]

Fortify Ends and Penetrations

> **Since people are by nature imitative and teachable, regularly they would show each other the effect of their buildings and their glorious inventions.... They understood rich material for building to be a lavish and abundant part of nature from which they drew; they nourished and increased it through skill and they equipped life with a delightful elegance.**
> —VITRUVIUS[12]

We have looked at how pattern derives from structural needs and how it must be woven to develop the strength of the built assembly. But pattern may also be woven by geometry. That is to say, we often use the geometry of shape to create interlocking forms. This type of geometric weaving typically relies on the mechanical engagement of the shape it abuts in the larger field. Considering simple pavements that transmit dispersing forces as we tread upon them, we can easily visualize how their patterns often rely on the interlocking shape of the paving module to stay woven. Different civilizations have worked to solve such dispersing forces in myriad ways. The Roman solution of carving the bottoms of their road stones into points and interlocking the pattern polygonally proved exceptionally durable, since the downward pressure of use on the road continually worked to bed the stones ever more firmly.[13] Combined with the interlocked polygonal shape, both solutions worked to fix the stones in place, keeping them from spreading. William Manchester points out that "in the year 1500, after a thousand years of neglect, the roads built by the Romans were still the best on the continent."[14]

The containment of forces within a stone building or the containment of spreading forces in areas paved by stone is one of the primary issues facing all builders of masonry and leads us inexorably to the next and third Rule of Bondwork: "fortify ends

The Roman roads combined two strategies to resist the natural spreading forces that all pavements face: the bottoms of the heavy stones are chiseled to a pencil point, causing the stones to set more deeply with use, and the roughly polygonal shapes key the stones together in a collective embrace. Ostia, Italy and penetrations." Entropy—unwinding and collapse—is a worthy adversary, and opposing it might be considered the builder's supra-personal task. This third Rule speaks directly to these spreading forces, and the strategy required to resist them. It specifically addresses areas of inherent structural weakness, and thus both our safety and the longevity of the result.

Concerning pavements, one simple solution is to add heavy (thicker) stones to the pattern edges to resist and contain forces that would cause the stone to drift. Adding weight and mass to the liminal edge is often combined with the strategy of interlocking pattern geometry. Until the mid-twentieth century, stone pavements were always bedded in some combination of rubble and sand. The rubble provides the road base or foundation, sand the leveling agent or setting bed. Once compacted and with the edges of the pavements constrained, additional sand or stone dust is swept into the open joints between the stones. Unless the ground has been excavated previously (making the subgrade prone to settlement), the work is likely secured and stable for multiple decades, until tree roots or groundwater work to undermine it.[15] Once that inevitably happens, the pavers are easily excavated and rebedded in additional sand, a simple and inexpensive process.

Of course, even with heavy stones such as curbs at the edge of our pavements, spreading forces remain at work, and we often encounter uneven paved areas. Our modern solution, contemporary with the wide adoption of high-strength Portland cement mortars, adheres the paving stones to a concrete substrate, or "sub-slab." On the face of it, this strategy is wildly successful. But that is only half the story. These new "rafts" of concrete and mortar-set pavers will also eventually fail and shift position. The ground plane is always moving, changing, through the action of such natural forces as frost, tree roots, earthquakes, and the effects of ground and surface water. The Jungian scholar Robert Johnson writes, "The vibrant energy of the earth inevitably triumphs over human efforts to suppress it, just as a tree's growing roots will eventually burst through a concrete sidewalk that has been laid over them."[16] Even when set in mortar, these larger sections will eventually be pushed out of alignment. And there's an additional difficulty that arises when we set our pavements in mortar, fixing them to a concrete sub-slab. If we discover that we need to make a change, removing the mortared stones can be difficult, and they often break in the process. The mortar, whose strength and adhesion to the stone makes it

LEFT
This contemporary application of basalt paving (called "bluestone" in Australia) relies on high-strength Portland cement–based mortar to affix the pavements to the concrete sub-slab. The method typically precludes the reuse of the material in the future, since the strength of this bond will likely exceed the strength of the stone, making recycling impractical. Melbourne, Australia

RIGHT
Large Roman stair blocks work to contain the spreading forces that all pavements face. Such pavements set in rubble and sand have remained in perfect position for two millennia, a reminder that some ancient methods are impossible to improve upon. Jerash, Jordan

capable of resisting the spreading forces, thus proves the project's undoing, often dooming the stone to the landfill, since the cost of labor to salvage the mortared stone exceeds its market value.

Although mortar-set paving on concrete sub-slab has become the de facto specification of our modern world, the impulse to attempt to fix stone in stasis was an illusion from the beginning. My intuition tells me that this yearning may be a holdover from the pioneer mentality of Manifest Destiny, when exploiting our natural world was considered not only an American right but part of our collective duty as citizens. This nineteenth-century philosophy seems to spring from a desire to dominate nature, to fix it firmly in place, and to support our illusion of control over the larger cosmos. But we would do well to remember the ancient wisdom of Heraclitus, who reminds us that "everything gives way and nothing stays fixed," perhaps the essential truth of our natural world and the things we build upon it.

In that light, we must reevaluate our intention to fix stone pavements irrevocably in mortar. As the resources of our world continue to be consumed, all of us must work to make sure our energy footprint is lighter and less thoughtless, our waste minimized. Even the concrete sub-slab for stone paving is not "free." Production and placement of concrete (and mortar) is estimated to account for 8.6 percent of greenhouse gases annually.[17] As the pages of this book have reminded us frequently, not all of our social progress is forward. Sometimes the old methods, such as setting stone pavements on compacted soil or rock, or sweeping

European cities have reused their pavements for centuries by bedding the larger stones in sand or weak mortar that may be easily removed when repairs or changes are required. In this example, the natural drift of the pavements is contained by bracketing them between the fixed positions of the masonry buildings bordering the street. Tarragona, Spain

The Khmer pavements are neither woven by geometry nor locked by heavy stone borders. In general, soil conditions were such that stone paving was minimally used, largely occurring only in courtyards and formal approaches to large temples. Angkor Wat, Cambodia

While these pavements may have been created out of a decorative impulse, their woven pattern arises from a deeper preoccupation with interlocking geometry that has concerned stonemasons since the beginning of architecture. Agra, India

the joints with sand or dust, do not lend themselves to improvement.

In modern usage, the bracketing of pavements with a thickened edge, even when the stone must be mortared to a concrete slab, remains a best practice. The thickened edge physically protects the thinner stone at the center. Since this center stone is usually only one and a half inches thick (the minimum recommended for exterior paving), the edges are fragile as well as visually weak. When we can both hide and protect this weakness, we have extended the work's longevity, aligning it within the five-thousand-year tradition of the Rules of Bondwork. Bracketing the pavements with the surrounding architecture also works well. Ringed with buildings, paved plazas are inherently stable, since the pavements have nowhere to move.

Sometimes the forces at work on a building are so extreme that every strategy available to masons must be called upon. Such was the case with the Eddystone Lighthouse, built between 1756 and 1759 by the British engineer John Smeaton. Working on a shoal of rock barely above the surface of the Atlantic off Plymouth, England, Smeaton used complex geometric weaving to fashion the tower's core shaft. Deploying tightly fitted granite blocks of considerable scale and mass to achieve maximum friction between the wrought faces, Smeaton pioneered an ingenious geometric weave that exploits the dovetail concept, mechanically interlocking each stone to the larger structure. The horizontal courses were then pinned with marble dowels to resist the shear produced by Atlantic storms.

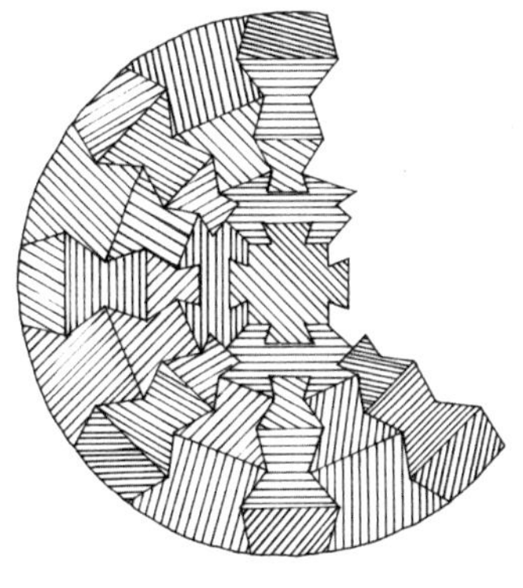

The strongest stone structure ever constructed is likely the Eddystone Lighthouse. It was designed and built by John Smeaton, who used large-scale granite blocks and innovative "dovetail bonding," the shapes interlocking to create a cohesive whole. Marble dowels connected the courses vertically. Cornwall, England

Furthermore, Smeaton created a waterproof lime mortar of his own design by working backward from the writings and recipes of Vitruvius. This hydraulic mortar is still in use today.[18] Of course, the mortar was not fashioned to "glue" the stones together but rather helped to waterproof the joints between the stone blocks. Considering the naturally low-strength mortars of the past, whose primary purpose was to "stop up the wind," there was never any risk that the mortar strength could exceed the stone strength and thereby damage it. Despite the lighthouse standing for more than 120 years and being one of the strongest stone structures that humankind had ever produced, the rocks supporting the foundation eventually began to split apart and undermined the structure.

OPPOSITE, CLOCKWISE FROM TOP LEFT
The pattern of this building, a quarry master's home, would be called "coursed rubble," since the joints are not finely wrought and the pattern runs in courses. Qufu, China

These finely wrought ashlar blocks were crafted using early iron tools. The face of the block has a fine texture called "point stalk," carved with the point chisel. Wushan, China

In isodomic ashlar, the heights of each course of stone are the same, even if the lengths of each block are random. The color of individual sandstone blocks has changed over centuries as distributed iron, originally carried by groundwater during the stone's formation, has reacted to the natural acidity of the rain. Nuremberg, Germany

In pseudisodomic masonry, the course height of the ashlar blocks alternates. Typically, the smaller course represents a bonder course, the width returning internally and working to tie the exterior stones to the interior hearting. Umm Qais, Jordan

Mortar and the Mating of Stones

Mortar, the material between stones, is often misunderstood. It goes without saying that mortar does not "glue" a wall together. That is a decidedly modern misunderstanding. Rather, what fills the spaces between stones—mortar, lime putty, common mud or earth, or even sheets of lead—acts only to stop the wind and help distribute the forces traveling down the wall through the various points of contact from stone to stone. This is particularly important in rubblework, where the mating between pieces is less ordered and thoughtful. Mortar-like materials added between the stones work in compression and, even if lacking strength themselves, transfer the forces from stone to stone, making a cohesive structural unit. This is why a brick chimney or hundred-year-old rubble wall continues to stand long after the mortar strength has dissipated. The fine grains of sand that composed the original mortar remain locked under compression, and as long as that compression remains constant, they continue to support the larger structure without complaint.

Re-pointing, also called tuck-pointing—the process of replacing old mortar—is required to maintain the water barrier only when the joints have been eroded by wind or frost. Mortar adds not strength but longevity, by re-shoring the weight transfer. Of course, by keeping water out of eroded areas, tuck-pointing inhibits future damage, since moisture in a masonry wall can weaken the material and structure, especially in areas where it can turn to ice.

The early Greeks used no mortar whatsoever, preferring to perfectly mate their stones. The preceding Egyptians, having invented the strategy of perfectly mated masonry, used a variety of clay and mud mixtures of minimal strength. As Mainstone explains, "Fitting the blocks together...distributes the stresses as uniformly as possible. Since there will only be compression to transmit, all that is required is a good uniform bearing between blocks."[19] When the careful dressing of the blocks is not performed, interposing a thin layer of material will transfer the compression regardless. Although this material may crack, Mainstone assures us that it is "unlikely to crush." These thin layers of primitive "mortar" serve as a lubricant in setting the blocks and improve the bearing and mating of the disparate surfaces. And, finally, as previously noted, mortar works to keep out damaging wind and driven rain or seawater.

The gold standard of isodomic ashlar is found at the Parthenon and represents a high point of both craftsmanship and design. The blocks are set entirely without mortar, being worked into full frictional contact. The four holes at the bottom of the photo indicate vandalism after the fall of Rome, when the internal iron cramps were chipped out of the standing walls, the scarcity of iron being so severe. Athens, Greece

TOP LEFT

This tightly wrought coursed ashlar maximizes the quarry yield by utilizing both random course heights as well as course lengths. The texture on the face of the material is called "split-faced." Giovinazzo, Italy

BOTTOM LEFT

The rectangular blocks of ashlar masonry may be the most basic and widely used pattern in stone architecture. Ashlar masonry is relatively easy to install, since the stacking of rectangular objects is easily taught. Montepulciano, Italy

TOP RIGHT

Random ashlar patterns are constructed of tightly fitting rectangles without continuous horizontal or vertical joints. This highly expressive example includes not only shapes mixed randomly but different textures and even a second stone type. Frankfurt, Germany

BOTTOM RIGHT

Isodomic ashlar is constructed of blocks of equal course height. Here, the block edge has been "rebated." The resulting shadow line adds further emphasis and formality to the texture and surface treatment. Wroclaw, Poland

This pattern of irregular coursed ashlar combines blocks that are not perfectly square. Nor do the courses run precisely horizontal. Yet the joints are tightly wrought, the defining characteristic of ashlar masonry. Jaisalmer, India

TOP LEFT

The dry-stone coursed rubble (set without mortar or mud) of Great Zimbabwe is in a class by itself. The walls of the Elliptical Building, measuring thirty-two feet tall, seventeen feet thick, and over eight hundred feet long, wind like a woven basket. Masvingo, Zimbabwe

BOTTOM LEFT

This example of coursed rubble has a foot on the threshold of polygonal masonry. The pattern remains bedded horizontally despite the random shapes of the stones. Lower Silesia, Poland

TOP RIGHT

At the temple of the Sphinx, the "re-entrant joint" in the center of this random ashlar illustrates a common method of working the stone in situ after its placement. Giza, Egypt

BOTTOM RIGHT

Often, building a stone structure is an iterative process completed over decades as resources permit. Reading this built graphic suggests the original foundation and building plinth were likely used as a livestock pen before the gables were added. Rizhao, China

TOP
The expressive work of this polygonal masonry in soft limestone is informed by the Rules of Bondwork but takes its own position. There are certainly stones resting on their points, contradicting the Rules, but these stones appear uniformly spaced, supporting a supposition that the masons were making stylistic choices rather than sporadic errors. Tarragona, Spain

BOTTOM
The Etruscan walls that encircle the hill town of ancient Falerii Veteres, Italy, were constructed in 300–400 BCE and demonstrate two key principles: the Cyclopean-scaled blocks give these walls their impenetrability and illustrate the principle of strength through mass, and the tightly wrought polygonal pattern literally weaves the wall together. Falerii Novi, Italy

This unusual wall is constructed with vertically oriented, split river stones. When fluvial stones are used, stone shims or "chinks" are often required to square off their round shape. More typically, we find stonework bedded horizontally for increased stability. This dramatic vertical orientation seems to reach upward, giving the wall a spirited exuberance. Shuitou, China

The trulli of southern Italy are constructed of limestone blocks in coursed rubble. Alberobello, Italy

The corbeled “shingle” found on the trulli of southern Italy can be described as dry-stone isodomic ashlar, since stone of similar course height is critical to its successful function. Alberobello, Italy

What we would consider "true mortar," a material having some inherent binding quality, was not used until the late Hellenic period and contained lime, coming either from burned seashells or burned limestone or marble.[20] We create lime by "calcining" (burning, typically in a kiln) limestone (CaCO3). Because the burning of freshly quarried stone, heavy with quarry sap, requires massive amounts of fuel, dried-out surface stone (dead stone) would be collected to be burned to achieve the required transformation, occurring at 900° C (1652° F). However, even less labor could be expended if the dead stone came from the demolition of older limestone buildings. Such scavenging of the built environment caused the loss of many exotic Roman marble slabs, brought at great expense from the far-flung reaches of the empire. These thin veneer stones, proving easy to pry loose and equally simple to transport, were looted and burned to create lime. Even more culturally devastating than the loss of rare colored marble was the almost complete destruction of the bas-relief carvings at Sahuré, Egypt. The German archaeologist Ludwig Borchardt, who excavated the Sahuré group of pyramids between 1902 and 1908, estimates that of more than ten thousand square meters of carved panels, "only about 150 square meters, broken into innumerable fragments, survived their depredations."[21] As it turned out, the fine Tura limestone from which the reliefs were carved produced the best lime.

The source of the burned lime aside, lime-based mortar has historically offered one of the best solutions for sealing the gaps and lubricating the setting of stone, the result of its workability. The mortar's texture is beautifully creamy and plastic, clay-like in its materiality. It is a joy to work with and remarkably waterproof. Regularly limewashed buildings, even when constructed of stone bedded with common mud for mortar, will last indefinitely.[22] Significantly, buildings covered in limewash and bathed in sunlight will also be cooler than non-limewashed buildings by 5 to 10° F (3 to 5° C).[23] Unfortunately, many stone buildings, particularly rural farmhouses, where the craftsmanship of stone assembly was inconsequential, have, in the architectural romanticism of the late twentieth century, had their limewashed render chiseled off, to detrimental effect. Without the protective limewash, driven rain quickly penetrates the stone walls and can undermine them. At a minimum, this moisture and damp penetrates to the interior, making the living space smell moldy and unhealthy. This destructive fad of removing the protective render

OPPOSITE
The fine Tura limestone, a favorite material of the Egyptian bas-relief carvers, proved an easy source of raw material for the lime kiln. It is estimated that from the Sahuré group of pyramids alone, more than one hundred thousand square feet of carved panels were burned to create lime for mortar and plaster by the civilizations that followed. Luxor, Egypt

and limewash remains pervasive and much more long-lasting than I could have predicted. I vividly recall Signor Fabrini, master mason of my Sienese guild, railing against this practice and its effects on the surrounding hill towns of Chianti in the early 1980s. Recent visits have confirmed that there is hardly a historical limewashed building to be found; the render and limewash have been detrimentally chiseled off to reveal the rustic stonework underneath.

Limestone's lack of crystalline structure means that a limestone pool terrace will be as much as 41° F (23° C) cooler underfoot than a terrace paved with a sandstone (such as bluestone) or granite.[24] I call this the "cool-deck" phenomenon, and it is a major factor influencing the material choice around swimming pools, where tender feet connect with scalding pavements.[25] This is the direct result of the crystal structure of non-limestone materials. Sandstone, granite, and concrete pavers all contain silica or quartz crystals, which capture the sun's light and hold it. Since light is heat, it is these crystals that burn our feet with their trapped energy.

Modern studies confirm what was considered common knowledge in my guild in Siena. Lime mortar is somehow

ABOVE
Limestone has many unusual and useful attributes, including its ability to simultaneously waterproof and cool a stone structure. Research has shown that a limewashed building in the sun will remain 5° F (3° C) cooler than without lime's protective coating. Locorotondo, Italy

"self-healing," meaning that it can "heal cracks and fissures... resulting in overall improvement of the mortar's durability."[26] This would certainly account for its success when deployed at the Eddystone Lighthouse, the high lime mortar closing tightly against the stone blocks to help the structure withstand the penetrating wind and spray from the storms of the English Channel.

Although many civilizations seized on lime as an ideal mortar base, the historical record reveals no comparable consensus on what lime might be mixed with to make mortar. The far Eastern builders of antiquity combined their lime with jaggery, a type of viscous sugar, to make a mortar paste, a practice still in use in the state of Karnataka, India.[27] Historically, the Khmer peoples mixed burned lime with palm sugar and liana juice to fuse both their early brick and (later) sandstone and laterite constructions.[28] Modern Peruvian builders mix their lime with cactus juice to make a "stickier mix."[29] The cathedral builders of Europe preferred lime putty, a type of lime paste that excluded sand. Since lime putty lacks body and is incapable of supporting the weight of the stone set upon it, balls or shims of lead were often used in the corners of the blocks to carry the stones' weight during construction—a method repeated for the modern construction of the National Cathedral in Washington, DC.[30] Lime putties rely on carbonization to set, meaning that they gain strength only as a result of contact with the atmosphere. Mainstone adds that this process happens "very slowly indeed in the centre of any large mass, and large proportions of the lime may remain unchanged even for thousands of years."[31]

This is not to say that the ancestral builders would not have liked to "glue" the stones together if they could have. From a contemporary twelfth-century Arab writer by the name of Edrisi who visited the great Lighthouse of Alexandria, we learn that "its separate layers of masonry are cemented together by molten lead, and this so firmly, that the whole is indissoluble, though the northern waves incessantly beat against it."[32] Lead has almost no strength and would be therefore useless as "cement" in the way Edrisi imagines. But it is an exceptional waterproofing method and is often used between coping stones to keep water from entering the wall below. In my own practice, I use lead wool to pack the joints of my outdoor stone tabletops, as the luster and finish perfectly complement the natural patina of the aging slabs. For the same reason, lead is typically used between the joints of fine stone tracery supporting the stained glass that characterizes Gothic cathedrals.

Regardless of whether the mortar joint is of lead or lime-based cement, the technique of closing a joint for weather has almost been lost. In modern practice, joints are commonly struck flush—leaving the joint flush and in plane with the wall face (or raked back), creating a recessed mortar joint, without the additional effort to work them closed and make a weathertight seal. However, with a little instruction, the technique of tooling a joint is easily taught. Working a joint smooth with an iron tool or bar will raise the finer particles of the mortar to the surface. This burnishing seals the gap between the stones and makes the joint watertight. Assuming that the mason works cleanly, the sponging of the joint may be avoided. Sponging the joint is a common error that washes away the mortar fines (the slurry of cement and lime that closes the joint to weather). Such work with a burnishing tool typically falls to an apprentice on the masonry team and is often the first assignment of real consequence. This work is important, since, as my guild teachers were quick to point out, jointing is a critical indication of quality and care. Bad jointing can make a great mason look poor, and good jointing can make an otherwise poor mason look good.

Woven Geometry

The type of internally woven geometry so successfully employed in the Eddystone Lighthouse proved a more common strategy than one would imagine. Such weavings are simply hard to see due to their hidden position within the structure of a wall. The principle can be more easily observed externally, when two stones have been interlocked to provide resistance to outside lateral forces that might overwhelm the assembly. Consider the interlocking "joggle" that works to keep an assembly aligned and is often used on curbstones and riverbank copings where, for example, the force of a barge could easily overwhelm the coping weight and knock it out of alignment. This joint, through simple interlocking geometry, works to add a modicum of strength to the assembly.

A similar strategy undergirds the dovetail or butterfly joint, apparently invented by the Egyptians. William Richard Lethaby cites J. E. Perring in reporting that at Dahshur, Egypt, many of the lower stones in the pyramid "were joined together by stone dovetails."[33] Cedar wedges survive at the Temple of Kom Ombo in Aswan, Egypt, whereas granite wedges were used to bind the sandstone in the many temples at Luxor.[34] Stone dovetails bridging a joint are designed to resist lateral pressure, real or

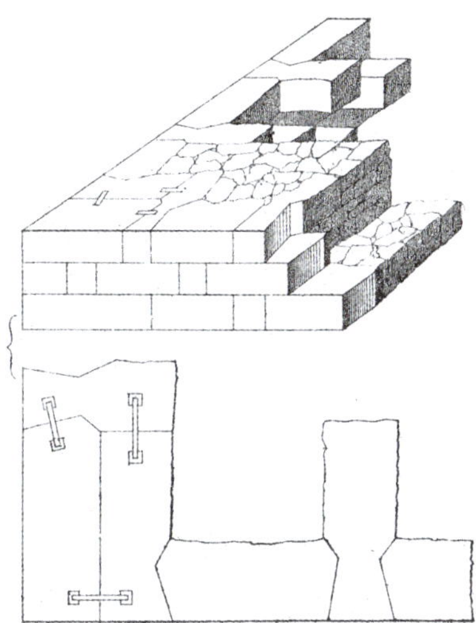

anticipated. In pharaonic Egypt, this technique was far from exact, and one sees wedges joining large blocks of stone at the bases of columns where no lateral pressure might ever exist. Almost three thousand years later, we find the dovetail or butterfly joints in the Khmer temple complexes of Cambodia, seemingly invented afresh.[35]

The early Iron Age brought the invention of the metal cramp, which, for the most part, quickly replaced the stone dovetail. Certainly, by the sixth century BCE, Greek builders relied heavily on them to construct the masterpiece that is the Parthenon.[36] In the medieval accounts of Winchester Castle in England, Malcolm Hislop cites the "making of fifty iron clamps to 'bind the tower.'"[37] We still use a variety of metal clamps to bind stonework, although working in modern thick stone veneers (two to eight inches in thickness), we are typically clipping back to structure rather than stone to stone. Today, to avoid the skill required for perfect encasement of the crimp in lead (if you get it wrong, it blows the stone apart through rust jacking), such attachments are almost always stainless steel, a simple but considerably more expensive modern alternative.

LEFT
The header and stretcher coursework of this bridge abutment leverages several strategies to build strength of assembly. Not only do the headers tie into the structural hearting of the interior, but the dovetail piece design keys the stretcher stone in place. The corners are also geometrically woven. Finally, metal clamps are added to mechanically lock the stones in.

RIGHT
Like so much of modern stonework, the Egyptians pioneered the use of "cramps" (or clamps) to hold stone together. Both cedar and stone were used extensively to bridge joints where the builders perceived lateral forces. Luxor, Egypt

ABOVE
Heavier blocks called quoins are used at corners to increase frictive strength. The method leverages the mass of the larger blocks to help counteract the inherent weakness of the wall termination; walls are naturally weak at their ends. Kyoto, Japan

Even more in buildings than in pavements, the third Rule—"fortify ends and penetrations"—finds traction. Although the primary force in a wall remains gravity, each time a window or door penetrates the building fabric, we divert the forces at play and introduce structural weaknesses. Even building corners are vulnerable, as they are weaker than sections of wall more central to the structure's mass. We may turn to heavier blocks at these exterior corners or at building penetrations, leveraging the block's increased mass and friction to resist the diverted forces. These larger blocks are called "quoins" and are woven into the wall pattern, returning left and right, alternating up the building corners.

Another fine example can be found in moat walls of medieval Japan. Here, large blocks (in full frictional contact) bolster the corners, adding strength to a vulnerability introduced by the required turn of the wall. The graceful shape created at the corner arises out of the vertical elongation of the battered form; the increased verticality at the top was added to better resist enemies who might climb the gentler slope of the wall's base. Although never the explicit purpose, these successive changes of pattern, texture, and scale alter and elevate the expression of

LEFT
The larger blocks forming the corners of these Japanese moat walls (in full frictive contact) bolster the strength of the form. Kyoto, Japan

RIGHT
These quoins are "expressed," meaning that through their chamfered edges and "proud" placement (that is, pronounced relative to the wall plane), they become a formal design element. Melbourne, Australia

the work. The result transcends the initial program and use. Eight hundred years later, we inherit a work of human labor that vibrates with visual honesty. Unlike the chicken and egg, we know which came first: structural and programmatic requirements drove design. In this way, good masonry is rarely strictly decorative. In stone design's best and highest form, changes of scale and texture are not applied out of whimsy but arise directly from structural concerns.

The same strategy may be seen where windows and doors introduce structural weakness through their necessary voids. Although typically smaller in scale than quoins used to secure the corner, the "rybat" (a type of small window quoin) serves the same purpose, bringing strength to the interruption of the larger pattern and buttressing the weakness caused by the penetration.

Including quoins and rybats in a wall changes the pattern and expression. What may have first seemed a stylistic convention in this light becomes the obvious response to a programmatically induced deficit: the weakness introduced by wall penetrations and ends. Quoins and rybats may be expressed in a stylized way, but they exist as an honest solution to a real structural issue. Arising from programmatic requirements and guided by the

TOP

Rybats—the cut stone terminations at the sides of windows and doorways—work to add strength at such penetrations, and perform the same function as their larger cousin, the "quoin," which occurs at the corners of buildings and walls. Melbourne, Australia

BOTTOM

These rybats are powerfully expressed, visually punching past the staid plane of the larger wall. The contrast between the smooth wall finish and the exaggerated rybats engages dynamic shadows cast by the strong sun of southern Italy. Lecce, Italy

LEFT
Larger frictive blocks placed at the corners (quoins) are particularly important in this case, to counteract the spreading forces of the herringbone pattern. Shuitou, China

RIGHT
This coping has been integrated into the larger architectural form of the wall. Yet notice the carved drip channel (also called "throating") tucked under and behind the leading edge of the molding. Cape Town, South Africa

Rules of Bondwork, the design response stays true. In this way, the design is not imposed upon the building but rather results from a response to structural necessity. This alignment between the requirement and the response gives rise to a visual harmony whose pitch-perfect resonance may be felt, if not fully explained.

The tops of walls are also vulnerable. Not only are they the typical point of entry for water—always a destructive force in masonry—but the hurly-burly of use, let alone enemy attack, exposes the top of a wall to actions that might unspool the woven strength of the larger assembly. For this reason, "fortify ends and penetrations" requires that heavy copings sit atop our walls. Like quoins, large capstones add downward pressure to the assembly, effectively driving the wall pattern into closer frictive embrace. Copings should be as long as possible to minimize the number of vertical joints where the pieces abut to limit water's access to the wall's interior. In the very best work, the vertical joints in the coping are sealed with lead wool. In all but the most arid climates, copings must be sloped a minimum of 5 percent so that water does not sit or pool on the capstone. The overhang should not be less than two and a half inches to allow for the placement of a drip channel, a timeless detail that works to shed the water and keep it from wicking underneath. Without a drip channel, the water will invariably enter the center of the wall cavity through the natural process of liquid adhesion.[38]

Chapter Five

FOUNDATION AND AESTHETICS

We cannot know what will come of the great teachings, we just have to let them flow unabated and not take fright if they suddenly rush into the natural ravines of life and vanish underground and race along unknowable channels.

—RAINER MARIA RILKE[1]

At least part of the "unknowable channels" Rainer Maria Rilke cites are found in the rock of the earth, similarly referenced by the infamous general Meng Tian, master builder of long sections of the Great Wall of China. When finally held to account for his brutality and the incomprehensible number of lives sacrificed during construction, he chose suicide over execution, confessing that he had "offended the cosmos itself by cutting through the veins of the earth, tampering with natural forces best left undisturbed."[2]

Within my own guild in Siena, the mystery of the Earth's lines and channels sparked much discussion, since they informed the places where one *might* build, and others where one dared not. Such discourse springs from the fourth Rule, "build on a good foundation," a concept so deeply ingrained in modern construction that we hardly pause to consider it. Indeed, modern geological science has removed much of the guesswork from our placement of architecture—or so it seems until our structures fail or shift catastrophically. Investigation may reveal shifting micro-conditions, a type of "vein of the earth" we tardily come to understand.

Sometimes the fault can be assigned to foundation failure. Since the foundational support of a building is generally hidden, a builder's fraud of shorting the specified materials often goes undetected until catastrophe strikes. Employees from my stone company working in Sichuan, China, in 2008 witnessed the aftermath of the massive 7.9 magnitude earthquake that toppled twenty-story apartment blocks as if they were small blocks of firewood standing on end, ready to be split. Reviewing their firsthand photographs moved all of us to deep distress. Investigation would prove that these buildings, constructed in rural townships outside the purview of city building inspectors, often lacked basic footings, or had concrete footings with undersized rebar or no steel reinforcement. The concrete itself was weak, too, shorted of Portland cement by unscrupulous builders and developers.[3]

As modern science has confirmed, building foundations can be explicitly counted on only when connected to living bedrock below. As we learned in chapter 3, bedrock is rock that remains connected deep in the earth with groundwater continuing to migrate through its mass. True bedrock can carry approximately one thousand tons per square meter in load, a fact intuitively understood by builders since time immemorial. As the gospel parables say: "And the rain descended, and the floods came, and the winds blew, and beat upon that house: and it fell not: for it was founded upon a rock."[4]

William Bryant Logan reminds us that "it is perilous to forget that buildings don't simply sit on the ground with the weightlessness of pen lines on paper; they rest on the very surface of the Earth."[5] This resting on the Earth holds more complexity than we might imagine, since the task of finding bedrock often proves difficult. For example, the search for bedrock to support the massive Cathedral of Saint John the Divine in New York nearly bankrupted the project, thanks to pockets of soft rock and underground springs.[6] Financier J. P. Morgan eventually stepped up with a blank check to complete the task. The excavation finally hit consolidated bedrock seventy-two feet below grade.[7]

Though a foundation seventy-two feet in depth sounds extreme, especially if you are one of the unfortunates charged with excavating the pit, such depths are supported by the Rule that "foundations should be one-third as deep as the structure is to be high." Most of the cathedrals of Europe took this advice to heart. Twenty years spent on a foundation was not uncommon, with footing depths of forty to sixty-five feet.

Adding to the complexity of the assignment, the bedrock we seek is rarely where we aspire to build, such that digging deep enough to find it is sometimes not an option. Frequently we are forced to compromise and find ways to support our structures in less-than-ideal conditions. Modern geologists speak of "regolith," the unconsolidated rock that typically supports our structures. The layperson might name it more simply: gravel, soil, or earth. To the credit of geologists, they are seeing soil in the context of deep time. Through their eyes and timescale, dirt is simply rock not fully formed; the vast forces of the cosmos are continually making the igneous, metamorphic, and sedimentary material of our planet.[8] In premodern times, being directed to build on dirt (or something worse) versus bedrock, where geology aligns to support heavy structures, might be seen as the first compromise of any master builder.

TOP
Fortifications often rise on the high points or liminal edges of alluvial plains. In addition to the increased vistas such locations present (always a benefit to see one's enemies approaching), the rise often indicates easy access to living rock, the ideal foundation for permanent structures. Erfoud, Morocco

BOTTOM
The quality of the "regolith" (unconsolidated rock) and soil that supports most structures has a direct correlation to the potential longevity of the structure. Uneven settlement will quickly topple a building, as gravity overwhelms the components no longer bedded horizontally. Ubud, Bali, Indonesia

TOP
So stable is natural bedrock as a foundation that no matter how violent their history, walls built upon bedrock almost always outlast the civilizations that built them. These are some of the oldest walls in Europe, dating from the time of Homer, roughly 1200 BCE. Mycenae, Greece

BOTTOM
This unusual example shows foundation stones for a staircase "scribed" to the living rock (when the abutting block is cut to fit the shape of the stone that proceeds it), rather than "keyed" into it to form the level base (when the previous stone would be carved or "keyed" to accept the stone that would rest upon it). The result is not as strong, but considering that it only supported the staircase and the weight of those entering the Odegal basadi temple complex, it was more a charming inconsistency than a catastrophic error. Shravanabelagola, India

Builders who fail to find bedrock to support their architecture have, throughout history, created some remarkable adaptations. The Rule to "build on a good foundation" contains the admonition that "stone supports wood," implying that wood does not support stone. Ironically, we discover some sophisticated solutions to building on regolith involving rafts of reeds or wood to float the structure upon the shifting "veins of the earth." Consider the ziggurat builders who learned to "plait layers of woven reed mats between the courses of brick, which made a simple, slender raft to spread and equalize the downward force of the temple and the spreading splash-response of the soil."[9] The archaeologist Henry Hodges adds that the "layers of reed matting were incorporated as the construction progressed. The matting served as a reinforcement in much the same way as steel rods do in a modern concrete building."[10] Modern foundation engineers use rafts of concrete to achieve stability in much the same way.

At York Minster in England, the massive continuous footings of the eleventh-century addition were constructed of a combination of stone interlaced with "stout oak timbers."[11] In fact, wood piers have been extensively employed in ancient footings. Saint Mark's Basilica in Venice and Saint Isaac's Cathedral in St. Petersburg used tens of thousands of them. The Venetian method of submerged wood piers, immersed in saltwater and thus deprived of oxygen that degrades the wood, is now well understood. Italian builders were hired by the czar to build Saint

LEFT
The Rule "stone supports wood" is almost universally followed by all cultures. The exceptions offer insight into foundation problems of great complexity. Kyoto, Japan

RIGHT
Although the Rule "stone supports wood" would seem to prohibit a wood lintel over a large door or opening, exceptions do occur. Typically, such a deviation is made only when the strength of the local stone is marginal, or a lintel block of sufficient length cannot be procured. Laon, France

Isaac's Cathedral in the marshes of St. Petersburg using the Venetian method. In both examples, the scale of the effort is hard to overstate. We know that for Saint Isaac's, whole forests were taken to provide the twenty-four thousand pinewood piles. These cylindrical tree trunks were tarred with bitumen and driven down to a depth of twenty feet to act, in effect, like modern steel "needle piles." Once the piles were submerged, a woven mass of granite blocks measuring twenty-three feet in thickness was built on top to bring the foundation to final grade. The cathedral then rises another 333 feet in height.[12]

Rowland J. Mainstone describes the sinking of a structure in its entirety as "inconvenient," whereas differential sinking is much more serious. If the sinking occurs "over the width of a single footing it will lead to the tilting (as in the notorious case of the Tower of Pisa) or the collapse of whatever is built on it.... The objective is not merely to keep the super-structure above ground, but to support it without undue movement."[13]

Of course, we rarely have the opportunity to excavate old footings unless they are failing. A startling example can be found

ABOVE
Built to lean intentionally as a garden folly, this small tower is more disconcerting than amusing. Viterbo, Italy

in the foundation support for the massive walls of Winchester Cathedral, which reach eighty feet in height. Since the subsoil contained a thick layer of peat, the foundation consisted of whole beech logs overlaid by an exceptionally weak concrete. Yet the assembly stood firm until 1905, when the lowering of the water table allowed the log foundation to decay.[14]

Although it may seem obvious in these days of concrete caissons and monolithic spread footings, it is useful to review the first principles of foundation. A foundation must be both "hard and solid," which means that it must be compressible and incompressible at the same rate throughout. The primary grain of the stone must be laid for maximum strength, perpendicular to the pressure coming from above. And finally, the foundation must be sufficiently deep to protect it from atmospheric changes. Although this typically refers to the heave that can force a structure up and out of alignment through the action of frost, it can also relate to keeping wood pilings sufficiently waterlogged (and therefore oxygen free) so as to maintain the core strength of the materials used.[15]

Here, too, we see how an understanding of differing material strengths becomes paramount. A foundation built of granite blocks can support forty-five to sixty tons per square foot, while a foundation of limestone blocks holds only sixteen tons. Medium-density limestone fares even more poorly, at eight tons per square foot.[16] To put this in context, a cathedral such as Saint Isaac's is said to weigh three hundred thousand tons. With such a vast difference between the program requirements and the marshy ground on which the czar ordered the cathedral constructed, it is a wonder of human ingenuity that the builders succeeded. Armed only with empirical Rules, the guild-trained designers created lasting solutions to foundational problems whose complexity would challenge modern builders equipped with gigabytes of geotechnical data.

Grounded Mass

Considering what we know about the frictional contact of the largest stones to create the strongest assemblies, it follows that the most massive blocks procured would be reserved for the foundation and base of a building, where the most critical strength is required. The inherent weakness introduced by joints is largely avoided with super-scaled materials. This strategy leads us to an important subset of this Rule that directly influences design. "Scale it appropriately to need" requires us to put

ABOVE
The Rule "scale it appropriately to need" leverages the frictive strength of massive stones to bring added strength where buildings are weakest. Thus, we often find the largest stones at the bottom of the wall to resist the immense weight bearing down upon them. St. Petersburg, Russia

the largest stones at the base of our structure, programmatically lightening our load as we progress upward. In practice, this was good for all, since the heaviest—strongest—stones were also the most difficult to move.

The mission of "scale it appropriately to need" invariably leads us to the next masonry Rule, "select each stone to its best use," since not every stone can be scaled to fit all programs. In this way, "select each stone to its best use" works to clarify how we resolve programmatic and structural requirements within the individual stone typology. Knowing the limited bearing capacity of medium-density limestone, "select each stone to its best use" would prohibit its usage for cathedral foundations, yet its specification for wall facings might be perfectly adequate. The same light-colored limestone would be a failure as a countertop (absorbing and reacting to food acids) or a pool deck (suffering erosion and staining by chlorine). Likewise, given the degradation of medium-density limestone by sulfuric acid in urban atmospheres, it makes little sense to specify this vulnerable material for buildings in cities, where concentrations of acidic pollutants are highest.

"Select each stone to its best use" requires that we have a clear understanding of the innate capacities of our building materials. Will a building stone be subject to freeze-thaw cycles that may weaken or degrade the natural planes of weakness that every stone contains? Will our material choice, in a specific location of the building, be adequate to fulfill the legacy promise that drove us to stone in the first place? Can blocks of sufficient length be procured to span the large penetrations and windows that modern architecture frequently requires? Such analysis is critical to our successful specification for stone architecture.

Foundational Illusions: Moving Past the Explicitly Rational

Returning to "scale it appropriately to need" and the example of stones diminishing in scale as the building or wall rises in height, we approach the Rule of Bondwork I found most surprising in my guild study. This Rule is concerned with equilibrium and the balancing of physical and "optical" stresses. As discussed, "scale it appropriately to need" implores us to put our largest stone shapes at the bottom of the wall. However, as we move upward, the natural lightening of the scale of stone gives us an added boon, the illusion of height. In the context of the Rules of Bondwork, this sleight of hand is called "make it sweet to the eye," and, with modern science diminishing the value of Rules like "build on a good foundation," "sweetness" may prove to have the most direct bearing on our consideration of twenty-first-century stone usage.

During my apprenticeship, the concept of "make it sweet to the eye" was conveyed to me on a particularly cold day as we huddled over our lunch and strong local wine. Signor Fabrini, the same master mason of my medieval guild who had eyed me so warily on the fateful day of my application, now took the end of his stick measure in hand, slowly drawing a circle in the construction dust on the floor between us. He finished chewing as he regarded me thoughtfully. I recall thinking something important was coming, but I had no inkling how powerfully his words would echo through the decades of my practice that followed. The room was quiet, and it seemed that even the distant birdsong and city noise stilled, bringing his words into sharper focus. Time elongated in that strange way I now recognize when one's experience comes to an intersection of some import. He started slowly and cryptically:

> You must remember, Riccardo, we once believed the Earth to be flat, so building and finding true horizontal once seemed not only obtainable but absolute. Our teachers understood the subtlety of this predicament, namely, that our eye plays tricks on our perception, and making something appear horizontal may in point of fact require that it be bent. That is to say, everything must work to *visually* support the primary axes of horizontal and vertical. The key word is "visual." It must look right. In this context, it is interesting to consider our recent knowledge: the Earth, in fact, is round. This helps reinforce what our guild has always known. For if the very surface of the Earth is curved, then even the concept of a level horizon—the basis of horizontal—is illusionary. "Make it sweet to the eye" speaks to that foundational illusion.[17]

At the time, what stood out foremost in Signor Fabrini's lesson was his reference to "our recent knowledge" concerning the roundness of the Earth: as if five hundred years ago was last week, and the periods of classical Rome, Greece, and Egypt were some small number of years before that. Although Signor Fabrini's larger lesson took decades to unspool, it remains a central tenet of design mastery as I have come to understand it. Stated in today's vernacular, "Even if you can prove it is level, if it looks wrong, you have failed." However, when you succeed, as articulated by the German artist Albrecht Dürer, you succeed beyond all expectation: "This looks good and carries well."[18]

"Make it sweet to the eye" has profound implications for design. It suggests we look deeply into the very basis of human perception, pressing for solutions to correct our innate visual limitations. The Rule points the path toward entasis (the visual correcting of columns) and speaks to some of the more dramatic contrivances we find on the Parthenon. Eugène Emmanuel Viollet-le-Duc noted that the ancient Greeks "found expression in a very close and subtle study of the natural laws and instinctive requirements of the eye."[19] In one of the best examples, the architrave on the Parthenon bows almost twenty inches end to end to correct for visual sagging. The architectural historian William Bell Dinsmoor quotes Percy Gardner: "The whole building is constructed, so to speak, on a subjective rather than an objective basis; it is intended not to be mathematically accurate, but to be adapted to the eye of the spectator."[20] Dinsmoor himself continues:

OPPOSITE
The Parthenon's builders heavily modified the dictates of strictly mechanical dimensioning. There were multiple accommodations to the essential geometry during its design and installation to correct for viewers' optical limitations. Athens, Greece

> The delicate curves and inclinations of the horizontal and vertical lines include the rising curves given to the stylobate and entablature in order to impart a feeling of life and to prevent the appearance of sagging, the convex curve to which the entasis of the columns was worked in order to correct the optical illusion of concavity which might have resulted if the sides had been straight, and the slight inward inclinations of the axes of the columns so as to give the whole building an appearance of greater strength; all entailed a mathematical precision in the setting out of the work and in its execution which is probably unparalleled in the world.[21]

The Parthenon's builders heavily modified the dictates of strictly mechanical dimensioning. There were multiple accommodations to the essential geometry during its design and installation to correct for the viewer's optical limitations. These adjustments stand at the heart of the Rule "make it sweet to the eye." The fact that this ancient temple of Athena still sings to us through the ages, some 2,400 years after its marble blocks were fitted in mortarless embrace, stands as testament to the power of the builder's sophisticated intuition.

The inescapable truth which we are forced to confront is that proportions in the final are felt, not measured. Returning yet again to the wisdom of Viollet-le-Duc, "The eye has an instinctive sense of what is reasonable...and why should not our eyes be possessed of a special instinct, as our mind is, which antecedently to any, can possess the sense of justice and injustice, of right and wrong."[22] How is it that we feel such visual injustice? Or, conversely, how is it that certain works stand out and seem to gather admiration over the centuries? How did Leonardo da Vinci's *Mona Lisa* and Michelangelo's *David* become the standard by which other artworks are measured? These touchstones appear to be pan-cultural, timeless in both their recognition and their appeal. Similarly, we can point to masonry masterpieces whose essential materiality, form, and function have achieved obvious transcendence. Although some works capture our imagination and remain important for their use of breakthrough technology or historical relevance sui generis, works like the Parthenon, Taj Mahal, and Pantheon are clearly in a class by themselves. This may be as close to universal beauty as humans are likely to achieve. The "how" of such a question lives at the center of "make it sweet to the eye" and no doubt will engage and energize our search for generations to come.

The design concept of symmetry is also approached by the Rule "make it sweet to the eye," though not in the way one might imagine. The Rule is not concerned with visual mirroring in the style of the parterre garden in fifteenth-century France; rather, its focus is the balancing of stresses. Although never the explicit goal, such structural balancing often results in visual balance and harmony. As such, it is part and parcel of the optical refinements that "make it sweet to the eye" works to address. Such visual symmetry rises from the primary goal of equilibrium, whereby the constructive system works oppositionally and offers active resistance to pressure.[23] Viollet-le-Duc offers: "It is certain that the architects of the Middle Ages had not the same idea respecting symmetry as the ancients.... What they sought for was rather the balancing of masses and details than their exact similarity."[24]

"Make it sweet to the eye" gives us the artistic freedom to move past the explicitly rational as the source and final limit of our expression. The Rule invites us to go deeper into the realm of our intuition—the hidden and unknowable source from which such works ultimately spring—encouraging us to follow the small voice of the soul in our journey. In this way, the Rule provides a bridge into sculpture (including architecture as sculpture), since, in the final analysis, only by navigating the uncomfortable distance between the rational and the intuitive self do we hold the possibility of creating work capable of transcending our specific time and place.

LEFT
The precise mirroring often found in later interpretations of classical architecture may arise from fad or fashion, but regardless, such symmetry often works in equilibrium to resolve the forces internal to the building. Cape Town, South Africa

RIGHT
Visual symmetry and equilibrium are the inseparable twin sisters of masonry architecture. Cape Town, South Africa

The most sophisticated examples of architectural symmetry incorporate geometric sleights of hand whereby dimensions may be stretched, platforms crowned, columns tapered, and the vagaries of human vision accommodated and corrected for. Bagnaia, Italy

The Rules offer us structure and method. They remind us that the essential materiality of stone and its relationship to gravity must inform its successful use. That said, the limitations of these ancient Rules are also obvious. Although humankind has looked for a blueprint or formula for universal beauty, such a perfect ratio does not exist. The harder work must still be done as we feel our way forward, stretching the math here, contracting the perspective there. In the final analysis, knowing when to bend and when to break rules remains central to the task of mastery. The Rules of Bondwork offer us signage toward the destination, but they never ensure our arrival. Rather, this journey is uniquely our own. It is a road we travel with our decades of practice and slow learning built on the foundation of the empirical wisdom that is our shared cultural inheritance.

ABOVE
The zipper path at Ryōan-ji weaves stone pavers to create a visual metaphor between the highly abstracted order of the composed garden and the chaos of the green world beyond. Kyoto, Japan

III.

A MANIFESTO

Chapter Six

SETTLED CONVICTIONS

MATERIAL RATIONALISM AND THE FUTURE OF STONE USE

If the soul is not served, the architecture has failed.
—THOMAS MOORE[1]

When we study the stone-built environment and work to read or decode the grammar of any masonry building or monument, we glimpse a snapshot of time and place. Our language acknowledges this. "Written in stone" remains a fitting coda to a finished, fixed, or unchangeable accomplishment. Preliterate cultures used the surfaces of their buildings to convey teachings and reinforce cultural values. In viewing the often-exaggerated styles of sculptural carving on the cathedrals of Europe, one might conclude that they were the graphic novels of the day, communicating complex metaphors to an illiterate public in reductive and stylized, highly charged grammar. The carved temple or cathedral, in fact, may be viewed as a book "revealing an esoteric teaching."[2]

In many instances, the European Freemasons surreptitiously encouraged pagan idolatry by incorporating archetypal motifs into structural elements. A symbol of wild forest energy, this ancient Green Man is carved into the projecting keystone supporting the engaged column above. Frankfurt, Germany

Reading our contemporary stone buildings may reveal painful truths concerning the vacuousness of modern life. I don't mean to "throw stones" at my colleagues and peers, but if the Rules of Bondwork offer a Rosetta Stone for the decoding of stone architecture, there are some pointed inquiries to be made and august professionals to be held to account. The Jungian psychologist James Hollis predicts that "from such cultural artifacts [we] might discern both the modernist achievement and disfigurement of soul."[3] Those remarks bring emphasis to the scope of our current crisis, for never has the use of stone been more widespread and more poorly executed. We have plainly reached a crossroads where the denial and diminishment of our intellectual and material nature have been allowed to compromise the expression of human life and its essential meaningfulness. With no material more worthy of our reappraisal, such poverty of expression adds urgency to our reassessment.

By neglecting our own story, or by telling that story so poorly in our architecture, have we not voted against our own legacy? Continuing to assemble buildings with twenty-year lifespans not only supports a destructive cultural norm where precious materials are wasted—quickly fated for the burgeoning landfill—but leaves our own chapter in the book of the built environment largely blank, defined primarily by an absence of excellent buildings. We find a poverty of architectural expression—buildings that lack material integrity, thoughtful craftsmanship, or meaningful visual punctuation of any sort. Most of what is built

today will soon be demolished, the cost of retrofitting far exceeding the apparent cost of wholesale replacement, and our modern built environment failing utterly to contribute a chapter to the book of the ages.[4]

Of course, each age seems to lament the current state of building. At the turn of the last century, William Richard Lethaby noted sharply:

> We have, as in the classic revival, compilers instead of artists, and machines instead of workmen, and, worst of all, a public that, constantly advancing in numbers, influence, and architectural ambition, is unable to distinguish good building work and true imaginative art from mechanical copying, scholastic imitation, whimsical conceit, and vulgar pretense.[5]

However, in working to identify current excellence, one finds the paucity of good contemporary work undeniable. Perhaps most interesting, the best of what has been built remains solidly within the language of traditional stone technique. While the program and forms of the new architecture may be inventive and fresh, the stone application is often more conventional than expected. There may be a crisp detail or a fresh interpretation to a single Rule of Bondwork, but overall, the advances are generally small and carefully considered. Yet stone's contribution to the modern project remains substantial, performing the bulk of the heavy lifting it always has. Although no longer carrying the forces of the structure, the masonry continues to ground the architecture to the earth, adding visual weight and substance. Stone always provides critical texture and tone, sometimes through pattern, other times by its absence. On the best projects, these vital elements are imperceptibly tuned, their expression finely balanced in support of the larger composition.

OPPOSITE
Pagan gargoyles serve important structural purposes, as they launch rainwater far from the building foundations. They also serve as a type of windbreak, slowing potentially destructive breezes as they wrap around a large structure in much the same way that projecting rocks will break the water speed of a river or stream.
Milan, Italy

The architect Kengo Kuma designed this museum, whose decisively modern pattern mixes solid and void with simple sawn strips of sandstone. The resulting horizontality references the traditional running bond pattern of the ancient rice barn behind. Nasu, Japan

This modern two-story townhouse designed by Studio Archea embodies both historical precedent and modern interpretation. Leffe, Italy

This structure by Fabrizio Rossi Prodi is clad in stone, but the material takes a back seat to the thoughtful ordering of the columns and careful tuning of the mass. Florence, Italy

Although the pattern and texture follow traditional techniques, the modern volumes and contemporary forms of the larger composition bring vitality and freshness to this design by Giovanni Michelucci. Campi Bisenzio, Italy

Raffaele Cavadi's design for a small town's city hall repeatedly inverts traditional expectations. Although the stonework breaks no new ground, it reads as texture, tying the building to the local geography and vernacular architecture. Iragna, Switzerland

The stone coursing of headers and stretchers offers a twist on the traditional imperative to set stones horizontally ("weave for strength"). Its contemporary feeling is reinforced by vertical orientation of the larger stretcher course, contradicting the Rule for fresh visual effect. Milan, Italy

Essential Principles of a New Stonework

Play versus Plug and Play

Although obvious, a simple insight of Rowland J. Mainstone is worth stating specifically: "Design must precede construction. The form must first be conceived as a unique spatial configuration."[6] Only when a design is fully developed can we hope to convey that design in drawings or models. Moving a concept from one's imagination, to paper, to 3D models and mock-ups is the real work of making buildings. The Dutch architect Lars Spuybroek reminds us that things "are made through meticulous drawing and redrawing, extensive discussions on the factory floor and in workshops, and the making of dozens of models and prototypes."[7]

Once design professionals separated themselves from the physical work of building, many lost touch with the essential physicality of manipulating materials into structure. In many instances, modern architecture has become merely the integration of various "plug and play" systems—for example, how a glazing and facade system integrates with structural and HVAC systems. The resulting container of these systems, the actual building, receives only abstract consideration, an aesthetic review divorced from the messier reality of the materials, real world, and specific site. As the results of our designed urban spaces frequently testify, those deeper reasons for a building's being—its civic and social contribution, its relationship and impact in all its multifaceted manifestations—often remain neglected. For stone in modern building to regain its expressive vocabulary, we must rediscover the incremental physical process of "tuning the form."

The empirical knowledge we carry becomes activated only when we engage the materials physically. One simple method of engagement is the development of the full-scale site mock-up. For those outside the profession, this is a section of the structure—built at full scale—showing critical material intersections and craftsmanship. Working at scale provides invaluable real-world insight into the building and improves every aspect of both the design and material integration, adding fresh air and vital oxygen to the intellectual hothouse that architecture has become. Physical engagement moves architecture out of one's head, out of the theoretical, and back into our hands, reuniting mind with body and spirit. Such engagement also goes some distance to heal the historical rift that has developed between physical and intellectual labor. By returning to the primary action of physical

OPPOSITE
Chinese architect Wen Shu salvaged materials from the neighborhood that was demolished and cleared to build this museum. An innovative pattern evolved from the haphazard material delivered to the masons laying the walls. Ningbo, China

manipulation, we return to the first principles from whence the rules of design spring. In the light and shadow of the building site, our vision for the project engages the physical world. We are reminded that natural materials weather over time, absorbing and repelling moisture at different rates, expanding and contracting in heat and cold. Outside the realm of paper and computer screens, we are once again builders, forming light and space, adjusting texture and tuning pattern. The architect Louis Kahn reminds us that "when you have all the answers about a building before you start building it, your answers are not true. The building gives you answers as it grows and becomes itself."[8]

To some outside the creative process, this may look and sound a lot like "play." Indeed, the manipulation of large blocks, stacking and unstacking to find a hidden order, is profoundly enjoyable. The psychologist Erik Erikson reminds us that this type of energy and exploration is "technical work on material objects."[9] Play, by this definition, remains an important part of creative discovery. Time must be spent to manipulate, tinker, and physically explore. It is a process without shortcuts, since working thoughtfully takes time.

Time and the Race to the Bottom

The Faustian bargain that modern architecture struck with technology has sped the design process by eliminating much repetition in the form of basic tasks. But the true cost of this has become clear. Our endless modern search for velocity and the false equation that "time equals money" has gutted the historical approach by which contemplation, incubation, and iteration paved the path to excellence. Faster design may be the conduit to greater wealth, but it is almost always the enemy of considered structure and nuanced expression. Although it is a minor example of how technology enables our race to the bottom, skipping the mock-up and models inhibits our ability to nurture the small voice of instinct, a critical ingredient in the process of creative assembly and expression. The contemporary sociologist Richard Sennett sums it up succinctly when he reminds us that "slow craft time also enables the work of reflection and imagination—which the push for quick results cannot."[10]

Empirical Knowledge

In addition to using mock-ups, a second step in our effort to reimagine the future of stone is the vital reintroduction of artisans to our design teams. Let us freely admit that the expressive stone walls of a cowshed rival our most expensive and dignified civic architecture by leading professionals. Top-tier education is not a guarantee of common sense, nor does it offer a monopoly on beauty or taste. Within my guild in Siena, design was considered an art, and said to be *solvitur ambulando* (Latin for "solved by walking"). The learned masters looked to results rather than to exalted processes of reasoning. We all know it is possible to reason admirably and yet produce dull and uninteresting designs, since in the end, proportions are felt, not measured.

Paper architecture, removed from physical materials and the innate sense that those materials possess, is unlikely to create the next masterpiece of our age. Even Leonardo da Vinci, writing at the end of the fifteenth century, cautioned: "To me it seems that those sciences are vain and full of errors which are not born of experience, the mother of all certainty, and which do not end in the experience observed, that is, whose origin or middle or end do not come to us through any of the five senses."[11] With this admonition in mind, let us include at our table those who have spent their lives in physical connection with materials. We need artisans to help us unlock the essential materiality of what we strive to build with, no matter their title. Sculptors, masons, quarry workers, brick and stone masons—all live lives in direct communication with the ancient energy of the Earth and have earned their hard-won knowledge. By including those who have earned material authority on our creative teams, we hope—at a minimum—to avoid imposing design that is disconnected from and unsupported by the material's core attributes.

Mind-Body Connection

In an ideal world, we would insist that our design professionals possess both hands-on expertise and scholarship. According to the scholar Pamela Long, Vitruvius came to the same conclusion: "Architects who possessed manual skills but no education, he insisted, could not achieve authority commensurate with their labors, while those who put their trust entirely in reasoning and letters follow a shadow rather than reality. Those who have mastered both are fully armed and have arrived more quickly and with authority at their goal."[12] Part of this problem is the way we prepare architects and designers for their professions. At a lecture I gave to the Royal Institute of Australian Architects in 2012, I learned that the prestigious University of Sydney requires only four hours of instruction on the entire field of masonry for its graduating students of architecture. And this in a city with a proud and well-developed tradition of stone buildings. Contrast the modern educational method with the training for architects at the turn of the twentieth century, which typically included physical experience in each of the traditional building materials. Aspiring architects gained a hands-on understanding of building components, the rationale being that if one has tried to carve a molding or waterproof a window penetration, one should be able to assess the constructability of that type of architectural detail. The talented Scottish dry-stone waller Neil Rippingale reminds us that "on paper, you can *draw* a man carrying an elephant. It's harder to do so in reality."[13] Insisting that the study of architecture include some time trying to carry elephants has been, and will continue to be, a good idea. Since architecture is fundamentally the integration of different materials, knowing how to build and add flashing to a stone wall, attach a roof truss, and waterproof a building penetration must remain basic knowledge for architecture professionals.

All of this works to point us to the core disciplines that mastery requires and how our approach may have slipped off track. Consider how teaching in the medieval guilds stands in opposition to the modern instruction in design. Even to this day, the manual craft disciplines, unlike the purely theoretical arts, retain the felt knowledge and somatic truths that manipulating materials provides. The cathedral builders created masterworks at least partly because they viscerally—within their very bodies and bones—understood the materials with which they worked. In such a paradigm, a work's success arises directly out of the larger connection of the full spirit. There is no bifurcation of the self, nothing shut off or shut down, in the conception or execution of the work. Rather, it is allowed to emerge naturally through physical engagement. Mind and body are engaged, working the problem as one.

Highest and Best Use

In one of the great moments of twentieth-century architecture, the US architect Louis Kahn, speaking to graduate students at the University of Pennsylvania, held out a brick and posed a simple but profound question: "What do you want, Brick?"[14] Of course, a brick is merely a human-made stone, and the master masons I worked under in Italy would have understood the question intrinsically, since it informed the essence of their daily practice. Hands-on practitioners of masonry around the

A brick is merely a human-made stone. When the US architect Louis Kahn famously posed the question, "What do you want, Brick?" he pointed to a central consideration that must inform stone's highest and best use. Kahn, for his part, insisted that a brick wants to be an arch. Luxor, Egypt world evaluate every stone they touch for its highest and best use, asking and answering this very question: What does the stone want? What is its desire? As we now know, evaluating grain direction and aligning that grain against the forces within the structure is key to good masonry. So is choosing the best face—burying the chipped and fractured surface. These are among the countless decisions involved in standard journeyman workmanship. At the most basic level, knowing intrinsic strengths and weaknesses of a material is central to knowing how to design and build with it.

Kahn's question, then, asks us to consider the essential qualities of stone in order to put it to its greatest use. Once the program for the building is defined, the material itself leads us to its design. In this way, our goal becomes clear: to discover the design that the stone dictates. When we can align our programmatic usage with the right material, we begin a promising design journey. Said more simply, our clarity of program (what we want the stone to do) leads us to the type of stone to employ (specification), pattern, and texture (strength), revealing, finally, our ultimate design.

When we support the authenticity and integrity of the material, each stone invariably will lead us to a certain type of usage. As Kahn sensed, the material's very essence contains its destiny. Carl Jung's metaphor that each acorn contains an oak tree speaks directly to my point. Although referring to human potential and transcendence, Jung reminds us that not every acorn fulfills its destiny and becomes a tree, asking that we consider just how many things must fall into perfect alignment before the acorn succeeds.[15] Applying the metaphor to stone, some material will invariably be sliced thin and mounted in private jets or yachts. Not every quarry block has the chance to become a voussoir in an arch or a pavement surrounding the Taj Mahal. Yet nearly every stone is capable of a higher mission. Part of our responsibility as sentient beings may be to align this material potential with a higher purpose of our making. If we harvest the stone for our use, why not deploy it to advance the larger cause of civilization rather than consume it thoughtlessly, discarding it once fad and fashion have announced its obsolescence?

As craftspeople, we anthropomorphize the materials we work with. This phenomenon stems from decades of physical engagement, projecting our consciousness into the material as a way of understanding what is needed in our search for clues on how the material will dictate the design. Consider the master wood-carver George Nakashima's insight:

> When trees mature, it is fair and moral that they are cut for man's use, as they would soon decay and return to the earth. Trees have a yearning to live again, perhaps to provide the beauty, strength and utility to serve man, even to become an object of great artistic worth.[16]

As Nakashima suggests, the material itself might object to being wasted when its yearning is to serve and provide beauty. Whether or not you trust his conclusion, what remains beyond debate is the implied call to action as we consider the deployment of the world's remaining resources. If slicing stone thinly leads us to soon enough throw it away like out-of-date wallpaper, then we have failed both the material and the generations yet unborn. Choosing to create banality with a material capable of transcendent beauty may not be a literal crime, but it offends the senses and dulls our world. Our goal must be to do better.

The Implicit Harm of the Purely Rational

Somewhere during the eighteenth-century Enlightenment, the cult of reason coalesced, becoming the rationalistic basis of modern education whereby we have come to give credence to only two realms of experience: reason and sense perception. Of course, it must be acknowledged that this new focus on logic gave us scientific reasoning (and the natural sciences), which has added immeasurably to our knowledge of the world. Yet the righteous certainty of the strictly rational has come at a profound cost, as the intermediary realm of imagination gradually faded into the background. Fewer and fewer craftspeople, artists, poets, and philosophers have been able to affirm "the structural coherence of imagination...against the tyranny of a technical and mechanical reason which threatened the imaginative basis of all human experience."[17] Such ultra-rationalism may even be the root cause of our drift away from the primal human instinct to make beautiful things. Consider Spuybroek's caution that the very concept of beauty has gone out of fashion:

> Nowadays, you would have to travel the planet to find a single artist claiming to be trying to create something beautiful, and probably that artist would lack talent or be extremely religious.... Beauty passed art by a century ago. The word

> describes a world of perfect forms, proportion, harmony, even order, a world of what Alberti called *cocinnitas*, transcendent, even divine laws that made things luminous.[18]

We are to stop trying to create something beautiful? Really? Conclusions of this sort litter the academic libraries of our contemporary world. Had the ivy-clad walls been able to contain them, they might prove easier to ignore. Such is not the case. Articulate scholars such as Spuybroek, possessed of broad academic platforms to distribute and popularize their dystopian views, give cover to those who exploit the profession for fame rather than soulful contribution. The predictable results are ill-considered vogues like Brutalist architecture and those who, like the US architect Frank Gehry, would have us consider asphalt and chain-link fence as interior finishes.[19] Although neither of these examples relates to stone per se, both are illustrative of the extreme positions society is willing to accept in the name of novelty. It suggests that when we lose track of beauty as an organizing principle, we are truly lost. Not only does our built environment suffer, but generations of young design professionals are programmed to disregard the lifelong effort involved in nurturing the small voice of the soul that might contribute something resplendent. The wholesale rejection of the "divine laws that made things luminous" might be laughable had its influence not been so widely embraced. Instead, we suffer the clichéd results in the modern cities we inhabit.

The seeds of this wholesale rejection of beauty were likely planted by the Austrian modernist architect Adolf Loos, who, writing at the beginning of the last century, published an influential essay under the revealing title "Ornament and Crime." Expounding his theory of radical aesthetic purism in the extraordinary cant of the day, Loos claimed that ornament inflicts serious injury on human health, on the national budget, and hence on cultural evolution.[20] He went on to explain that to decorate a building with anything "pretty" was a sin against the true profession of the architect, which he now defined in purely functional terms.[21] Le Corbusier soon added his considerable weight to the nascent movement, claiming that only undecorated buildings were "honest," since the true goal was only to create watertight, functional structures. And so it goes.

As proposed, putting craftspeople back at the design table might go some distance toward grounding our efforts to create

projects worth the material and physical effort their construction requires. With our climate irrevocably warming, these concerns are mission critical to our survival as a species. No craft master whose life has been spent working to transform physical materials is interested in wasting natural resources for fad or fashion. This respect has been earned through decades of working to coax natural materials into vitality. Such a life's work informs a key tenet of mastery: that what "matters is living up to a standard of excellence inherent in their craft."[22]

Material Rationalism

Imagining what form stonework might take in the coming centuries brings us to the nature of aesthetics and to the central questions of art and nature, beauty and the artificial, that have occupied architecture from the beginning. Although the question is a moving target, each period strives to answer it through the filter of its moment and concerns. The question remains difficult to pin down—indeed, impossible to solve, except in a preliminary way. This doesn't make the searching questions futile, only more urgent. Consider the pleasure and power informing our gaze at Chartres Cathedral in France, and how those conclusions were

ABOVE
The Dutch architect Lars Spuybroek reminds us that "beauty once described a world of perfect forms, proportion, harmony, even order, a world of what Alberti called concinnitas, transcendent, even divine laws that made things luminous." [Lars Spuybroek, The Sympathy of Things: Ruskin and Ecology of Design (London: Bloomsbury Academic, 2016; repr., 2017), 160. Bagnaia, Italy

turned on their head five hundred years later by Antoni Gaudí's Sagrada Familia in Barcelona. Indeed, with both structures containing such fully developed styles, who are we to say which is more pure, or more salient to the civilizations that follow?[23] Part of our modern task, then, must be to integrate these two extremes while adding our fresh response to the continuing dialogue.

How the target moves is a new question of some moment. Can such a shift be anticipated? It may prove easier to know when something has become worn out and needs to be discarded than to predict the future expression of those who remain unborn. Trying to see the future, we become aware of how poorly we see our present. Alain de Botton reminds us how the distance of time helps us regard objects or buildings without our current biases: "With the passage of time, we can gaze at a seventeenth-century statuette of the Virgin Mary untroubled by images of overzealous Jesuits or the fires of the Inquisition."[24]

It is useful to consider the philosophy underlying an aesthetic to better understand the expression that follows. Erwin Panofsky suggests that the humanistic mind demanded harmony: "Impeccable diction in writing, impeccable proportion in architecture." He contrasts that with the modern scholastic mind that insists "upon a gratuitous clarification of function through form just as it accepted and insisted upon a gratuitous clarification of thought through language."[25] Using this model, we can track the important shift from beauty and visual balance to a focus on revealing the structure. Alas, as often becomes the case, such visual logic has been taken much too far, as if any type of illusion is suspect. Through this modern lens, one could argue that the simple wearing of clothes becomes an untruthful adornment of the body. This view is a decidedly modern concern. Looking to history, we can track how illusion in architecture, veneers in particular, became vilified. Consider the Roman method of building structures with cost-effective materials and skinning them with thinner, rarer, and more beautiful materials. As Sennett puts it, "The building appeared to be made of something that in substance it was not; its materiality was disguised."[26] Recall from chapter I Eugène Emmanuel Viollet-le-Duc's vociferous complaint that Roman edifices received "a decoration of marble which has no absolutely necessary connection with that edifice; but the decoration is a kind of second structure, whose richness does not belie the material used nor the manner of using it." To his mind, the decorative stones topping the comely brick and rough-hewn stone of the structure lacked not only richness, but honesty. Such an illusion, he writes, constitutes "a sin against taste, for taste consists essentially in making the appearance of reality."[27]

This line of reasoning is foundational to the modern aesthetic and forms the basis for the elaborate praise the flying buttress continues to receive in nearly every text on cathedral architecture. One such typical assessment offered by Gerhard Rosenberg holds that the structure of the buttress "more than in any other member is the creative power of the functional principle apparent in its form."[28] Although undoubtedly an engineering breakthrough and justly iconic in its universal adoption during the period, I don't find the flying buttress particularly compelling as a form. To incorporate the buttress within the larger visual massing would seem an option worthy of investigation.[29] We don't necessarily need to make every connection explicit. Contrary to many modern practitioners, I find the false ceiling that hides the electrical conduit and air ducts a welcome civility when it simplifies and orders the visual information of an interior space.

Panofsky's insights help explain how we have come to find ourselves at such a painful architectural crossroads that veneers are allowed to float visually unsupported above grade, and modern fireplaces whirl with fans to assist the draw, burning fake clay logs. Once early modernist logic defined a fireplace as an "appliance" and a home as a "machine for living in," disguising the newly defined object became visually untruthful. But this confuses truth with plausibility and, more important to our purposes, may point to the end of an era, offering opportunity for new forms to rise afresh from muddled logic and worn-out traditions.

Ironically, a closer reading of Viollet-le-Duc's "making the appearance of reality" may point the way toward the future. Despite what modern engineering has made abundantly clear—that structural masonry is no longer safe for human habitation—the Rules of Bondwork still offer a path forward. Even if we can no longer build structural stonework, even if structure no longer requires stone to support it, stone remains visually imperative to tie a structure to the earth and give it visual weight. At a minimum, stone must continue to speak to us of safety and permanence. Even if only ceremonially, stone will remain foundational to our relationship to both the earth and the built environment we construct with it. In this way, it provides

The Getty Center, designed by the US architect Richard Meier, applies stone in unsettling ways. A hallmark of unconsidered masonry is stone floating in the air, as if by magic. Los Angeles, California

the "appearance of reality" through the illusion that it continues to do the heavy lifting of supporting structure. Quoting Viollet-le-Duc once again, "The skeleton is always manifest below the muscles of the body."[30] *Implying* the heavy lifting might be enough. Using thinner stone to save the material resources, freight, and labor aligns modern practice with resource scarcity and concern for the carbon footprint—foundational issues of our time. Designing our stonework in ways that appear structural, tying it to past uses and the concerns of our collective unconscious, must be central to whatever form our new aesthetic takes. Of course, realigning such concerns with essential excellence of effort and intention will go a long way toward making sure that the contribution is additive to the five-thousand-year tradition. "Doing our best" each day is materially different from dedicating our efforts toward "doing our fastest" and wholly divorced from an intention to "do it our cheapest."

Although I have nothing against illusion in the sense of modern *l'art pour l'art* aesthetics, I remain uninterested in using products that ape the real thing. I don't need concrete to be stamped and dyed, or tile to be inkjet printed and textured to look like stone. I find even more offensive the modern trend of cutting real stone into tiny modules and repeating small-scale patterns to compete with fake or "manufactured" stone.[31] It is illusion enough when honest material can be manipulated to give the impression of structure. For example, back-cutting solid blocks of stone to use as quoins or rybats gives the impression that the veneer is composed of much thicker stone and that the walls are structural. Within this modest sleight of hand, I find myself enthusiastic.

In that sense, my vision for the future might be called material rationalism, since the illusion is functional in intent. In service to function, illusion's goal is not deception per se but rather alignment with beauty, tradition, and the expression of stone's essential nature. The architecture critic Lewis Mumford reminds us that "our task is not to imitate the past, but to understand it, so that we may face the opportunit[ies] of our own day and deal with them in an equally creative spirit."[32] To that end, let us be the first of the twenty-first century to acknowledge the changes to be made. The structural use of stone as well as the traditional carbon-heavy process of modern building contributing to the warming of our planet must be modified. We should consider this settled law and move our considerable energy toward solutions. The answers, yet to be codified, will surely not

TOP
This fine example of rubblework and brick constructed as a veneer is Roman, and has been criticized by some academics for supporting "illusion" by hiding the humbler materials of the structure. The pattern is called opus reticulatum. Ostia, Italy

BOTTOM
This modern stone veneer mirrors the Roman method of building a structure with cost-effective materials skinned with thinner, rarer, more beautiful materials. Amman, Jordan

follow the current vogue in architecture, abandoning the visual logic of gravity—or beauty for that matter—as the essential organizing principle. Until gravity is *actually* abandoned, and our earthborn home is shed, such a position lacks basic common sense, no matter how modern we aspire to be. To that end, let us resolve to engage our creativity for solving how nonstructural stone can visually tie a structure to the earth without abandoning our history. Let us align our solutions with our architectural patrimony, creatively solving for our time and generation.

The Crushing Cost of Planned Obsolescence

Honoring the past traditions of our built environment should not be optional, for the past informs the future in ways we cannot fully articulate. T. S. Eliot concluded that our only superiority to the past derives from our capacity to include it as part of our present.[33] The writer William Faulkner took it a step further when he expanded: "The past is never dead. It's not even past."[34] In the same spirit, let us fight the construction of disposable buildings. There is nothing inexpensive when a building is made to last only a single generation. Even using allegedly "low-cost" materials, we pay a price too high when the building's remains are carted off to the landfill after only modest use. By insisting on higher-quality materials, we set the expectation for multicentury service, the true life cycle cost becoming *de minimis*. In this way, letting buildings evolve and be modified becomes part and parcel of how we carry the past forward, creating a palimpsest written with the materials and forms that have preceded us. Historical buildings should not be frozen monuments, prisons of past forms and styles that must never be altered. Surely there is a place for both preserving the best examples while encouraging creative alteration to meet changing needs. Flexible and innovative adaptations must be made the first instinct, not the last.

Charge It with Emotion

Once we commit to taking the time to express ourselves through our building efforts, we come to the larger question of life force and the quality and emotional charge of such engagement. One of our goals for the future use of stone must be to enhance the connection to deeper meaning that stone has always carried for humankind. Scholars distinguish between Apollonian and Dionysian, between serene (rational) and passionate architecture.[35] Certainly, serene architecture has its place. Consider essential utilities or select government ministries (the Internal

Revenue Service comes to mind), where the mission, although critical to civil society, may not align with passionate support of the program. Serene, unadorned utility can be soothing. In our modern world, high-arching freeway overpasses capture this Apollonian sentiment, where the simple accomplishment of the task, undecorated and unembellished, offers a restrained beauty.

The ancient Roman aqueducts, with their tightly organized rows of arches, present another case in point. The stone is roughly squared, not fussy, and the arches, composed of simple half circles, maximized efficiency, since the voussoir was mass-produced at the same shape and scale for each arch. This offers function first, with beauty arising out of alignment and proximity rather than a fine-tuning of form. We celebrate the aqueduct's achievement as structure, its attraction coming largely from utility, mass, and context. Even so, base utilitarian infrastructure still requires aesthetic consideration. Aristotle's passionate plea that the (Apollonian) "fortifications of a city should contribute to its adornment" still echoes.[36]

For all else—schools, airports, libraries, museums, public and private housing, religious buildings—let us strive for passionate expression of Dionysian architecture. Why should we settle for less? Even commercial enterprises and headquarters, reflecting the vigor of their capitalist effort, deserve what Lethaby called an architecture "touched with emotion."[37] Many who labor in design and building harbor this desire. Nonetheless, it is a tall order to organize disparate materials into something that elicits emotion. Our modern legal strictures make this particularly difficult. The primary focus of architecture and construction firms today has become, all too often, "not being sued" rather than working to make structure touched with emotion. Obviously, solving for the basic concerns of building safely remain tantamount. Lethaby, writing in England in 1911, reminds us that even in his time, "the great architectural question of to-day is how to build common damp-proof walls; simple solid floors; and above all, roofs better than the thin slate lids we are accustomed to."[38] This essential requirement stands as a low bar that we can certainly hurdle. At the same time, let us hold up the inner energy that seeks something larger from us. Hollis adds that "all of us, at all levels of collective participation, need to recover our nerve, to find the will and courage to ask more about meaning, to oppose depersonalization, to again stand up for soul."[39]

Acknowledging the difficulty of inspiring emotion through our constructions does not absolve us of failing to try. Our

Although this example is found in India, not Greece, Aristotle would surely have approved, since he lobbied tirelessly for beauty, insisting that even the city's battlements should contribute to its elegance. Shravanabelagola, India historical building record is replete with architectural triumphs that still thrill, structures that continue to move us long after the masters who assembled them are gone. What is it, then, that helps charge a space and make a building's sum mean more than its disparate parts?

Modern vocabulary falters as we approach such subjects. T. S. Eliot reminds us that

> Words strain,
> Crack and sometimes break, under the burden,
> Under the tension, slip, slide, perish,
> Decay with imprecision, will not stay in place,
> Will not stay still.[40]

The cult of rationalism may have successfully banished the vocabulary needed to approach subjects not offering easily measured results. Consider that scientific work is routinely presented as the only path to a knowledge of reality outside ourselves. Yet this completely ignores gnosis—the immediate, unmediated, phenomenological experience that a great building or any great work of art may provide.[41] For such discussions, we are forced to turn to modern depth psychology, quantum physics, and astronomy, where the inexplicable is posited with great seriousness.[42] At such moments it helps me to remember that just because science has yet to identify the unmistakable twinkle in a person's eye, this does nothing to negate its existence. Borrowing a term from depth psychology, we discover "numinosity"—an aura of awe and mystery usually associated with religious feeling, and a term we must now apply to the architecture of the future.[43] Hollis expands this principle and connects it with soul: "When Mystery winks at us, we realize that soul is not only in us, but also in the outer world.... We have been brought into life equipped to be the carriers of soul, recipients of soul and builders of soul."[44]

To right this modern wrong, we must have the courage to feel our way forward beyond known laws, working to reintegrate soul and ecstatic joy into our building program. We must thread a careful line between robust skepticism and open-mindedness, holding out the possibility that we have more to learn, more to discover. One simple way we might invigorate the Dionysian attributes in our architectural expression is to reintroduce handcrafted, hand-altered natural materials into our building programs. As a stone carver, I have a strong sense that a little of my artisan soul remains in each work I shape. As an artist, I feel it as clearly as I feel the hammer in my hand. As a writer and scholar, I can only search for the words to describe what I know, looking for support in what others have written. As a modern-day citizen whose culture insists on a single rationalist paradigm, I can only suggest that our science has yet to determine a way to measure what I know to be true: that craft is a manifestation of numinosity.

John Ruskin hinted at the divine presence of a thing—the numen—when he hailed the imperfection of craft as a sign of its quality, linking soulfulness to the material crafted by hand and acknowledging its superiority to that which is machined.[45] Although his concept is abstract, I invite you to let yourself feel the soulful difference between what has been wrought and what has been mechanically processed. As a child, I was fascinated by the special bent wooden spoon that seemed never to leave my grandmother's right hand. Later, turning over a worn hammer handle often used by my father, I marveled at how the battered wood still carried the vitality of its purpose. My love for the owners of these objects could not entirely explain their gravitational pull—a pull I now feel to be the residual energy of an object's use.[46] Spuybroek reminds us that "all relations between things are felt relations."[47] The concept may not be as esoteric as it sounds. Just listen to someone describe their first visit to Europe and the feeling of walking on ancient pavements or climbing the worn and dipped, hand-carved stone staircase between galleries at the Louvre. The energy of these experiences is palpable and reveals itself in the telling.

Ruskin's statement must be expanded. Yes, quality is one measure we care about. Machines have advanced and now produce high-quality objects and materials, so using diamond saws to slab our stair blocks makes perfect economic sense. But we must recognize that the machine's lack of soul will likely never allow the ephemeral quality that handcrafted objects possess. Viewed in this light, adding the imperfection of manual labor, perhaps by chopping or texturing those machined blocks with a hand axe, we have psychically warmed the material, activated and imbued it with the soul of the artisan. Like Ruskin, I have come to see the imperfection of the object as reflecting the imperfection of our own humanity. It may be our flaws that make us—and the materials we have wrought—valuable and perfect.

Giving context to Ruskin's time, Sennett reminds us that "against the rigorous perfection of the machine, the craftsman became an emblem of human individuality, this emblem composed concretely by the positive value placed on variations, flaws, and irregularities in handwork." He goes on to correctly identify the machine as the greatest dilemma faced by the modern artist-craftsperson and to remind us that in the Victorian era, "for the first time, the sheer quantity of uniform objects aroused concerns that the number would dull the senses, the uniform perfection of machined goods issuing no sympathetic invitation, no personal response."[48] Although it is impossible to pinpoint the cause, something has dulled our modern perceptions and stripped the soul from much of our material world. Perhaps this Victorian fear was not unfounded.

Yet the dilemma for the modern craftsperson arises in the identification of which machines can be used to improve our productivity without stripping the numen. Surfacing a block of stone with a bush hammer swung from a wooden handle or one still held by a human but driven by pneumatics is, I assert, immaterial, the difference largely ergonomic, not aesthetic. The resulting soulful variation of the human holding the tool comes through regardless. Nakashima observed: "As much as man controls the end product, there is no disadvantage in the use of modern machinery and there is no need for embarrassment.... A power plane can do in a few minutes what might require a day or more by hand. In a creative craft, it becomes a question of responsibility, whether it is man or the machine that controls the work's progress."[49]

When it comes to the surface of a finely crafted wood tabletop, the ephemeral signs of craft are likely to be sanded or planed away, unwanted. Whether one achieves that smooth surface by hand or machine is the choice of the artist-craftsperson, since only they will know if the tool is augmenting or diminishing the residual life force. This may be wholly different from stone, where such imperfections and maker's marks help to convey the soul of the carver.[50] Holding these two materials in contrast reinforces my view that each medium has its own rules that guide its ultimate expression. Only through decades spent in physical communion are we able to read this expression with confidence and learn to intuit whether the tool is adding or subtracting from the imbued energy of the result. In this light, I propose a new definition of mastery as "an accumulation of authority beyond scientific measure."

New technologies need to be relied upon for what they can do well. As time-saving devices to accurately rough out form, surely they have great value; as robots replacing the subtlety of the human expression, never. The subtlety of direct human expression through stone will never be matched, for the human "imperfection" of a hand-textured surface is the very thing that makes it exciting and vital. Ultimately, it is the human interaction with stone that lifts it from its natural state and gives it new voice, new notes of meaning, beyond the power of a machine. The central issue remains that machines lack soul.

Algorithms programmed to create artificial imperfection have already flooded our world, adding "human touch" to industrial products from "stonewashed" jeans to greeting cards. The results are wildly inconsistent, since machines are designed to ape a finished result rather than to reflect the journey traveled at the hand of the carver or craftsperson. In the artisanal model, the material's final result records the actual journey, not a counterfeit impression, hollow and without meaning. Machine-carved moldings in stone offer a poignant example. Not only does machine carving lack undercutting (a current technical limitation), but the result is dry and somehow brittle, even sterile. Instead, we find form without the human variation intrinsic to the hammer and chisel blow. Rather than extending and enhancing the architectural gems of the past, the new technology simply distorts. Assuming that technology continues to improve—and that these deadening attributes are ultimately able to be addressed—it remains unimaginable that the machine will ever be able to energetically charge the result the way human carving invariably does. Without soul itself, the machine can only mimic human imperfection. As it turns out, such a deficiency is not a math problem. If we could program soul, we would add it directly to the machine and, as so many science fiction plots predict, soon enough be working for the machine itself.

Modern machining has succeeded in adding new finishes to our range of options. The question remains if such finishes are expressive. Consider how advances in industrial diamond technology used for polishing now assure our ability to gloss virtually anything, such that what once seemed natural to stone is no longer a concern. Incredibly, this polishing even extends to sandstone, reflecting our image back to us like Narcissus at his pond. Spuybroek adds, "We live in a world where more things than ever surround us, but when we look at them, we see only ourselves, as in polished, reflective objects."[51] The Japanese writer Jun'ichirō Tanizaki reflects my own bias when he so eloquently states, "We do not dislike everything that shines, but we do prefer a pensive luster to a shallow brilliance, a murky light that, whether in a stone or an artifact, bespeaks a sheen of antiquity."[52]

So far, much of the bravura of modern processing has stripped stone of its elemental beauty, its unique expressiveness. Our ability to slice, dice, manipulate, and polish the material has, in many instances, visually plasticized it. Widespread mechanization has unleashed a cascade of unforeseen consequences. Significant reductions in material thickness have fundamentally changed the aesthetics of our use in far-reaching ways. The very solidity of stone, that foundational quality that often drove our selection in the first place, has in many instances been lost. The visual thinning of stone—its diminution in mass—has reduced it to another surface finish without conveying the sense of structure or wider mission. The modern language of stone reflects this profound change. Stone walls are now spoken of as a veneer, a cladding, a wallpaper, and are detailed as such. Mitered or ship-lapped exterior corners, or thin treads on staircases, are but a few of the ways that modern usage has damaged stone's core attributes. In many cases, the essential materiality of stone, the very basis of our original human attraction, has been undone.

In the face of such changes, modern stone architecture has become a combination of "products": off-the-shelf sizes and shapes applied to exterior surfaces, loudly proclaiming that the stonework for the building has been only lightly considered, if at all. Not truly designed, the stone application will certainly lack proportionality by demonstrating a one-size-fits-all mentality—a painful turn of events. That said, the convenience of assembling "products" cannot be overstated. With labor costs continuing to rise throughout industrialized nations, pre-crafted systems ready for installation are perhaps inevitable. Systems themselves are not the issue. Arguably, much of the mass affluence of the late twentieth century was a result of the success of our broader manufacturing systemization. Our mission, then, must be to guide modern stone design within the efficiencies of systemization, leveraging what is useful and discarding that which undermines stone's core attributes.

By including mystery and sacredness, working to embody them in an aesthetic construction, we move beyond "beauty" in an artful sense, creating the possibility of a highly charged, cathectic experience. In centuries past, knowing that your neighbor's hands cut the stone, chiseled the texture, and placed the

TOP
Although we are clearly anthropomorphizing when we ascribe to materials a "soul," it is not difficult to imagine a little of the soul of the artisan, perhaps the residual energy of human craft or use, remaining and affecting our perception. Angkor Wat, Cambodia

BOTTOM
This innovative finish at Kunsthalle Würth was created mechanically. The stone was split on an angle, not 90 degrees, creating a strong shadow line below each course. Schwäbisch Hall, Germany

ברלין
BERLIN
WESERMÜNDE
NORDEN
LEER
EMDEN
ESENS
JEVER
VAREL
AURICH
PAPENBURG
WILHELMSHAVEN
WEENER
HOPPSTÄDTEN
OLDENBURG
DELMENHORST
OBERSTEIN
MEPPEN
BREMEN
OSNABRÜCK
AUMUND-VEGESACK
LEMGO
SALZUFLEN
DETMOLD
NORTHEIM
HOLZMINDEN
GÖTTINGEN
HILDESHEIM
HAMELN
ELDAGSEN
PEINE
PATTENSEN
BÜCKEBURG
BARSINGHAUSEN
WUNSTORF
BRAUNSCHWEIG
VERDEN
WOLFENBÜTTEL
HANNOVER
LÜNEBURG
CELLE
HARBURG-WILHELMSBURG
ALTONA
WANDSBEK
KIEL
LÜBECK
ELMSHORN
BAD-SEGEBERG
SCHWERIN
GÜSTROW
FRIEDRICHSTADT
ROSTOCK

blocks into meaningful form made the emotional resonance of the materials and resulting structure entirely tangible. In modern society, the neighbors cutting the stone are likely much farther away, a product of global labor arbitrage. Still, the craftsperson's soul remains present. An essential humanity remains intact whether that stone was carved in India, China, or the suburb of my own US city. Furthermore, we value the human effort and skill required—a proficiency that engages both body and mind—and hold it meaningful, the finished object no longer a common rock but a work of human beings. As ever, the backstory informs the material's soulfulness and expands its emotional resonance.[53]

Historically, the embodied sanctity of meaningful construction would have been plain. In many projects that have endured, sacred aesthetics—godly and sacred beauty—were always the goal and inseparable from any building made and dedicated to God. The activated energy perceived in handcrafted materials supported the story of the work being of God, sanctified by God's hand in the work, its resonant mysteriousness aiding perception of the mystical. According to this narrative, humans could not have created the work alone, even if only inspired by God. God's existence supported the (divine) inspiration required to manifest the structure.[54] Although this line of reasoning may have run its course and fail to inspire our secular citizenry, it remains informative. Let's hope that we have grown more comfortable in sitting with such mystery and not just too numb to feel it. Our task then remains to apply it to our current structures, drawing out these complex lessons in support of creating "architecture as sanctuary for art, for culture, for enduring rituals, for the human spirit itself."[55]

As a design strategy, the incorporation of handcrafted or "activated" materials is not new. Suburban churches in the United States, often assembled with middling design, if any at all, will add a hand-hewn baptistery, baptismal font, or carved front doors to bring humanity to the T1-11 plywood or synthetic stucco siding. The strategy is helpful, yet raises the larger question of what else might be done. In the near past, with the universal adoption of the industrial sawmill, rough-sawn wood timbers replaced the hand-hewn hardwood beam, the soulful standard of previous centuries. Despite the loss of hand processing, the essential honesty and massing of the solid wood carried the day. Sawn blocks of stone with a natural or machine cleft face share the same ethos. Although no longer hand pitched into fine ashlars, the natural face and texture of the stone augments the

OPPOSITE
Although the roughly quarried stone blocks are sawn square and largely undressed, their massing and material honesty carry the memorial's gravitas. Rather than the crisp arris indicative of hand processing, we find drill marks and other signs of rough extraction, perhaps left as metaphor for the cities and towns torn from history by Nazi terror. Jerusalem, Israel

Homer recognized the power and mass of these ancient walls when he called the citadel "mighty walled Tiryns." These Cyclopean masonry walls originally measured nine or ten meters high and as much as seventeen meters in thickness, and are chinked rather than closely fitted. Argolis, Greece composition and supports the essential righteousness of the material.

Relying on the craftsmanship of assembly without charging and changing the individual pieces is also possible. The Cyclopean walls of Tiryns in the northeastern Peloponnese, Greece, retain their resonant vitality despite the wall builder's inability to shape the hard boulders themselves. The ancient fortifications of Japan work in similar ways. Lightly wrought—if at all—the thoughtful walls seem almost to vibrate with the intense craftsmanship of their assemblage.

In addition to being constructed of soulful, handcrafted materials, ancient spaces may have achieved much of their charged energy through the performance of ritual and offerings of sacrifice. With these now largely absent from modern life, I look to my modest family cabin for comparison. Our simple wood structure has been seasoned by the decades of laughter and wholesome meals shared, the joy of our family life palpable and obvious to visitors, who often comment on the unmistakable feeling the modest cottage engenders. Similarly, I have noticed the equivalent timbre when visiting the once personal spaces of powerful consciousness, be that Saint Francis of Assisi's tiny monastic cell at Cortona, Italy, or Nelson Mandela's prison cell at Robben Island, South Africa. Both unmistakably resonate with the spirit of their previous inhabitants.

Stone and brick spaces carry the energetic freight of humanity in a way that steel and glass structures seem to lack. Even with sparse visitorship during early morning or late-night hours, the endless flow of humanity through New York's Grand Central Station demonstrates the activated energy of transitional spaces. The building positively vibrates with the residual energy that works to psychically enliven the enormous masonry volume. Future science may one day prove that since masonry and wood materials are harvested directly from the earth, they have a closer affinity to the human soul than "second-generation" materials like steel and glass. Although composed of earth elements, steel and glass remain disconnected. Despite how quickly handcrafted expressions in metal or glass overcome such concern, the assembled spaces of mechanically extruded materials feel largely cold and hollow, the "why" flickering at the vanishing point of human comprehension.

My call for Dionysian architecture and emotionally charged materials may also be heard as a call to the grandeur of human aspiration, perhaps even a call to sculpture. If we strive to work at the intersection of the mind and poetic sensibility, perhaps the transition from building for shelter to building as sculpture will be natural and, in some ways, inevitable. Sculptors, working as they do to create energized symbols, know firsthand that virtually any object can be energized to communicate ideas and values. De Botton reminds us what talented sculptors teach us: that the big ideas can be "communicated in chunks of wood and string, or in plaster and metal contraptions, as well as they can in words or in human or animal likenesses."[56]

In large measure, we choose the depth of our own lives and the quantity of richness we draw from the world around us. By neglecting the hard-won empirical wisdom of the civilizations that preceded us, we shortchange our engagement and self-limit the vitality of our work.[57] The choice remains ours. Mystery, gnostic revelation, and the experience of the numinous have guided human civilization from the beginning and remain available to us. Let us reassess our contemporary values and reassert the expression of the human soul as the highest realm of human achievement.[58]

As the stone expression of the last five thousand years so elegantly testifies, these are not new ideas. They simply ring unfamiliar to senses deadened by thoughtless consumption, endless novelty, and disregard for the wisdom that our past contains. The Inca civilization made the leap to sculpture, as did the Egyptians before them. Yet the power of these ancient accomplishments so shatters our helpless modern image of self that we attribute their work to aliens, devising conspiracy theories in an attempt to reconcile their massive achievement against our own poverty of purpose.[59] We stutter in awe trying to describe how humans, working by hand, assembled the largest stone building/tomb/sculpture the world has ever seen, with tolerances measured to one-fiftieth of an inch. Or we consider the Inca craftspeople, who—without metal tools—collectively channeled compositions and stone patterns that the Japanese American sculptor Isamu Noguchi might aspire to create. As moderns, working almost exclusively in our heads and relying on the machine to limit our physical engagement at every turn, we cannot conceive how these "primitive societies" made so much of their short lives. So comfortable have we become in our failure that the very notion of working to create beauty has become passé, a conceit we may have forfeited at the altar of strict rationalism and science, Jung's dire warning long forgotten: "The more the critical reason dominates, the more impoverished life becomes."[60]

In closing, it is critical that we raise our own expectations, rejecting what Hollis calls the "small lives we live compared to our summons."[61] Dedicating oneself to unlocking the sympathy of the natural world—perhaps stone in particular—offers access to incomprehensible questions of existence. Ideally, these worthy questions will bring us the formative energy to vitally engage with life, to transcend the ordinary, and to combat our modern disregard for the numinous. There is no time to waste. Our world's finite resources are at the tipping point, their continued squander no longer an option. For the generations yet unborn, let us engage natural stone at the highest level, investing our human vitality as a soulful expression of our place in the cosmic timeline.

Richard Rhodes on location in 2017 at the Stone Museum designed by Kengo Kuma & Associates. The vast majority of photos in this book were taken in medium-format transparency with the Rolleiflex camera shown in the image. Nasu, Japan

Acknowledgments

My friend and mentor of thirty-eight years, John C. Coldewey, edited the first four chapters and helped me for more than a decade. His untimely death deprived us of completing the journey together, yet his gentle voice became my own as I labored on to see the book to completion. John taught me to write through patient explanation and humor. He is deeply missed.

A special thanks to Amy Gansell, who helped me locate much of the early research while juggling work on her dissertation and commute to Cambridge. Ashley Esarey was instrumental in helping me access the Avery Architectural and Fine Arts Library at Columbia University.

Additionally, this book has benefited from the time and expertise of the generous friends and colleagues who read my manuscript. Collectively, they have offered insight, corrections, and guidance on where my explanations fell short, or where my understanding was imperfect. Dealing as they were with a first-time author, this was yeoman's work, and I remain deeply appreciative of their effort. They did their very best, and any remaining flaws are entirely my own. My deep gratitude goes to:

Jeff Arnold
Fredrick W. Atherton
Daniele Balleri
Stephen W. J. Boyle
Douglas Bryant
Patty Burkart
Mary Catlin
Tamara Caulkins
Kevin Chester
William Elliot
Melissa Evans
Rebeca Firestone
Tom Flynn
David Gedye
Brad Goldberg
Nicola Golden
Saul Golden
Viva Hardigg
James Hollis
Mark Lange
David Laskin
Rachel Lavengood
Vincent R. Lee
Jeffrey R. Matz
Michael Mayes
Ryan McCaffrey
Ken Meyer
Reno Pisano
Richard L. Rhodes
Stephan Sullivan
Edward Tufte
Robert Valenti
Daniel Zatz

Glossary

alluvial: Referring to a rich soil of unconsolidated rock created by action of water but then deposited in a non-marine setting. See also *fluvial*.

anathyrosis: When abutting two finely wrought pieces of stone, the saving of time by recessing the interior edges of the block so that the abutting faces are not perfectly mated. Typically, this entails leaving a flat border around the perimeter of each stone to be joined. Conceived by the ancient Egyptians and utilized extensively by the Greeks, the technique is still in use today. In the ancient world, it can be commonly seen on abutting stones at the vertical joint where there is no frictional pressure. When full frictional contact will contribute significantly to the strength of the assembly, the technique should optimally not be used on top and bottom surfaces. Also known as "slack jointing."

anisotropic: Describing stone with a grain structure. Stone is typically stronger perpendicular to its grain structure. See also *grain* and *isotropic*.

aqueous: One of the three chief classes of stone (the others are *igneous* and *metamorphic*). Sedimentary by definition, aqueous rocks have been deposited or formed in water or air and include most limestones and sandstones.

architrave: Ornamental carved moldings surrounding doorways, windows, and other building penetrations; can also refer to the spanning member or beam between stone columns, or the first course of a classical *entablature* composed of two additional elements, the *frieze* and the *cornice*.

archivolt: A curved band of carving on the *voussoir* stones of an arch. The band typically follows the *extrados*, or outermost radius of the arch stones. This was originally the same as an architrave.

arris: The edge of a stone where it changes direction from the face to some other plane; also used to describe the scalloped edge of a fluted column.

ashlar: Finely wrought blocks of regular rectangles with tightly fitting joints (typically one-eighth inch in thickness or less).

ashlar pattern: A pattern composed of squares and rectangles of differing sizes fitted with joints of one-eighth-inch thickness or less.

baluster: One of a series of turned or carved members supporting a *coping* or handrail to form a balustrade.

barrel vault: A continuous ceiling of arched vaulting that is either semicircular or semi-elliptical.

base: One of three elements of a column (see *shaft* and *capital*), not to be confused with a pedestal, which supports the column's base. Adding confusion, the base can also occur at the bottom of a pedestal or *parapet* as one of the three elements that define those: *base* or plinth, *dado* or die, and cap or *cornice*.

bas-relief: From the French for "low-relief," a carving on a flat panel in which the projection from the surrounding surface is slight (see *proud*) and no part of the modeled form is undercut.

batter: A receding, sloped wall, often used for retaining walls or assemblies requiring great strength.

bed: In sedimentary stone such as sandstone and limestone, the primary grain or layering; also used to describe a layer of mortar ready to receive stone.

bed mold: The template for the bottom of a wrought stone piece.

bedrock: Living rock that remains sufficiently connected to the earth that groundwater still flows through it. It is the ideal foundation for structures of virtually any scale.

bevel: The slope formed when cutting off a corner at an angle. See also *chamfer*.

boasting: A method of dressing the face of a stone. The texture is created with a wide chisel, and the finished result is a series of fine lines in succession, often diagonal.

bond: The practice of defining piece size and shape to which the whole is bound into one compact mass. Additionally, it is a measure of how well woven an assembly of pieces is integrated into the greater fabric of pattern and structure. Each stone must be supported by a minimum of two stones below to be considered bonded.

bonder: An individual stone that runs from the wall face to the interior *hearting*, thus working to tie the wall facing to the interior of the structure. Sometimes called a "dead-head." In dry stone walling, this piece should extend at least two-thirds through the wall thickness, although it will often extend fully from one wall face through to the other.

bondwork: The practice of defining piece design to bind the assembly of pieces into the compact whole of the larger structure.

boss: A protruding *keystone* at a vaulting rib intersection.

bracket: A stone support projecting from the face of a wall. Typically ornamental, it rarely supports heavy loads. See also *corbel*.

buttress: A massing of stone in front of a wall to resist the horizontal forces that want to overturn it.

calcareous: Containing lime ($CaCO_3$).

calcining: Burning or roasting in a kiln—a critical part of the process for converting limestone to lime, one of the key ingredients in mortar. Lime is also calcined to make cement.

camber: A small upward curve given to a flat arch or long-spanning *lintel* to combat the visual illusion of sagging.

cantilever: An overhanging member extending some distance beyond its point of support.

capillary action: For stone, the wicking or drawing up of water into vertical stonework. This happens when the water's adhesion to the wall is stronger than the cohesive forces among the water molecules.

capital: One of the three elements of a column, this one sitting at the top; see also *base* and *shaft*. These are often elaborately carved or stylistically identified by the classical Greek orders: Doric, Ionic, and Corinthian. The term is restricted to columns and pilasters only.

caryatid: A figurative form, typically female, appearing to support the weight of spanning members such as *architraves* or *lintels*. These can be engaged with the wall or freestanding, taking the place of columns.

catenary: The curve of a chain or heavy wire while hanging, ends spaced apart. This natural shape offers nearly perfect compression when used as the line of an arch.

centering: The temporary mold for the support of an arch during the building process.

chamfer: The surface obtained by cutting off a 90-degree edge at 45 degrees. This treatment typically ends before any intersection or abutment, creating what is termed a "stop chamfer."

chinking: The process of adding smaller stones to shim or close the gaps of two adjoining pieces. These stones relieve the need to work the stone shapes into perfect contact and are often left exposed, becoming part of the larger pattern.

colonnade: A row of columns supporting an *entablature*.

cool deck: A phenomenon specific to limestone paving whereby, thanks to a lack of crystalline structure, which prevents it from trapping light (and thus heat), the surface remains 10 to 15 degrees cooler than other types of stones such as sandstone, granite, and even pavers made of concrete.

coping: The capstone of a wall, *parapet*, or *gable*. It generally (although not always) extends over the wall upon which it sits and is sloped to shed water. The portion overhanging the vertical wall face should include a *drip channel* carved on the underside. The term is also used to describe a thickened paving stone at a pool's edge.

corbel: A stone that projects outward from the wall surface, typically supporting a heavy load such as a roof truss. See also *bracket*. This term is often used as a verb, referring to the act of overhanging a stone or brick from the one below. In extreme cases, this overhang can be as much as 30 percent of the piece width.

cornice: A projecting stone at the top of a wall or *entablature*, often elaborately carved with architectural detail. The function is both to weight the top of the wall and to project so as to direct water away from the structure below. See also *architrave* and *frieze*. It can also describe the molding at the top of a wall where the wall intersects the ceiling.

course: A row of stone, often set to a specific height, typically running horizontally the length of a wall.

coursed rubble: Non-tightly-wrought stone running in horizontal *courses*.

cramp: A metal or slate geometric key or clip used to connect two adjoining stones; sometimes called a "dog" or "clamp." Iron cramps were often cased in lead to keep the metal from corroding and exploding the stone. See *rust jacking*.

crown: The highest part of an arch or vault.

curtain wall: The outermost wall of a fortification, connecting tower, bastion, et cetera.

Cyclopean masonry: Stonework of huge piece size as originally found in Mycenaean architecture. According to Greek myth, stone walls with individual pieces of epic scale were thought to have been built by a race of giants called Cyclopes.

dado: The body of a pedestal or *parapet*, between its cap and base, or the lower part of an interior wall when it is differentiated or decorated for contrast with the upper part of the same wall. Also called the "die."

damp coursing: A through-wall layer of flashing (often lead) that inhibits water from wicking upward into the wall's greater mass.

dead stone: Rock that has become unmoored, upthrust, or quarried whose internal *quarry sap* or groundwater has dried out.

dentils: Small repeating rectangular blocks carved ornamentally in the *bed mold* of a *cornice*, first appearing in Greek and Roman architecture and now a standard of the classical vernacular. It has been suggested that the word comes from "teeth," as it resembles a row of teeth.

dew point: The temperature at which condensation (water droplets) begins to form. This varies according to atmospheric pressure and humidity.

dimension stone: A designation for material that is quarried and shaped rather than collected or gathered.

dip: Within a particular quarry, the resting angle of the earth's stone strata.

dolmen: A simple Neolithic construction found primarily in northern Europe, with at least two stones supporting a horizontal slab, or a stone chamber. Slightly later versions can be found in North Africa, the Middle East, and Southeast Asia.

drafted margin: A decorative textured border added around the perimeter of a stone facing. The border is often applied to the forward faces of split-faced blocks to highlight the alignment of the *arris*, adding formality and refining contrast between the rough and the smooth.

draught: In carving, a longitudinal groove carved into a stone to indicate the depth of the overall reduction of the surface. When the groove is more formalized and finished, it is sometimes called a *rebate* or "rabbet."

dressing: The process of adding texture to the face of a stone.

drip: A small projection or molding designed to clear water from a building. See *drip channel*.

drip channel, drip molding: A carved or cut joint on the underside of a stone or masonry molding to discharge rainwater; sometimes called a *throating*.

efflorescence: The buildup of salts and minerals on new masonry faces through water migration from the substructure or material itself, typically, the result of not keeping the materials dry and covered during construction. It may also indicate water entering the wall or structure post-construction.

elastomeric form liner: An inexpensive rubber mold used to form concrete. At best, it can create abstract patterns that catch light and shadow, breaking up flat wall planes. At worst, the technology apes stonelike patterns, dulling our visual perceptions.

engaged column: A column that projects from a wall while remaining attached to it. The projection must be greater than half of the column's normal thickness; otherwise, it is considered a *pilaster*.

entablature: Sitting atop a series of columns, it is composed of three distinct elements: *architrave*, *frieze*, and *cornice*.

entasis: The slight swelling curve of a column to counteract the illusion that the sides are concave instead of straight.

extrados: The outer radius of an arch. See also *intrados*.

extradosed arch: An arch composed of *voussoirs* that resolve into the surrounding coursework. See also *non-extradosed*.

extruded joint: A mortar joint whose profile rises above and beyond the *arris* of the abutting stones.

face: The exposed surface of a stone or brick.

face-bedded: Describing stone set with the primary grain running vertically rather than horizontally, compromising the material's strength.

feather: A half-round of metal used to line a drilled hole in stone, allowing a "plug" or metal wedge to be driven into the hole, splitting the stone apart. Typically used in pairs (one on each side of the drilled hole).

ferrous: Containing iron. A ferrous clamp (also called a "cramp") must be encased in lead to prevent it from rusting and degrading. A non-ferrous clamp (for example, one made of stainless steel) would not require encasement.

finial: A decorative stone at the top of a structure or spire.

flagstone: Naturally occurring flat stones used for paving.

fluting: The scalloped edges of a decorative column—usually concave vertical channels that run in series around the perimeter of the shaft. They are typically cut into the column in situ, after installation.

fluvial: Stone shapes whose rounded forms were created or reshaped by water, usually by the action of a river or stream. See also *alluvial*.

flying buttress: A type of stone strut, connected and supported by an arch, built to resist horizontal forces of a much larger wall, often found on Gothic cathedrals to counteract thrust from the roof.

freestone: Stone without strong grain orientation (see *isotropic*). Such a stone can be readily worked in all directions, eliminating the difficulty of carving against the grain.

frieze: A decoratively carved band of stone supported by columns. This is the center section of the *entablature*, which is composed of two additional elements (see *architrave* and *cornice*). An uncarved version is called a *pediment*.

gable: A triangular end of a pitched roof.

galleting: Small spalls of stone or brick that are pushed into wet joints during construction. The purpose may be decorative or structural (particularly when brick or terra cotta is used), since absorbent masonry can help extract the moisture from the mortar, speeding its cure.

gargoyle: A stone spout, typically carved in fanciful animal forms, installed on a building face to launch draining rainwater far from the building.

Gothic arch: An arch pointed at its apex, where the two sides of the *intrados* meet.

grain: In building stone, the aligned composite particles from which the stone is composed, offering planes of weakness that may be exploited for extraction or shaping. Not all stone has useful grain structures, and this helps distinguish workable stone from types considered intractable. See also *isotropic* and *anisotropic*.

granite: A metamorphic stone primarily made of quartz, and one of the four primary types of building stone (the others are *limestone*/marble, *sandstone*, and slate).

grazing: The collecting of surface boulders or rocks for building.

grotto: A shady, often cave-like garden feature.

hammerstone: A harder stone used to shape a softer stone by pounding.

header: The end of a block of stone, laid perpendicular to the face of the wall and working to bind the face stones to the backing. If quarried and processed correctly, this is also the location of the tertiary or end grain on a block of stone. Also called the "perpend," a term in opposition to *stretcher*.

head joint: A vertical space between two abutting stones, which may or may not contain mortar.

hearting: The structural center of a wall, typically created with broken or random bits of stone or brick, set intentionally and tightly fitted for strength.

herringbone: A stone pattern using rectangles where the corner of the first rectangle starts at 45 degrees to the horizon. Subsequent courses abut this first course at 90 degrees, creating a zigzag pattern.

hydraulic mortar: A special lime-based mortar designed to set and harden underwater.

hysteresis: Describing, in true marbles, a small increase in volume after each exposure to heat or frost, which causes permanent deformation that eventually can break the clips holding the stones, causing them to fail and fall. The phenomenon is likely named for the time lag between cause and effect, the more common (non-stone) definition of this word.

igneous: Rock formed from molten lava or magma, for instance, granite or basalt; one of the three main geological classifications (the others are *metamorphic* and *aqueous*).

intrados: The inner radius of an arch. See also *extrados*.

irregular: Stone or pattern whose sides and angles are unequal in appearance.

isodomic: Course heights of the same dimension in a wall (piece lengths may vary), often used to describe fine ashlar.

isotropic: A classification of stone without obvious grain or planes of weakness.

jack arch: A flat *lintel* created from wedge-shaped *voussoir* stones.

jamb: The sides of a door, window, or other opening in a wall, often carved. It may also refer to the side walls of a fireplace surround.

joggle: A method of mechanically keying two stones together through interlocking geometry.

joggle joints: Square, semicircular, or angular shaped tongue-and-groove joints of equal depth carved into abutting stones. Sometimes the stones have only grooves on both abutting faces, and the void created is filled with lead. Another method is to fill this void with small pebbles, then pour in a slurry of pure cement.

joint: In the quarry, the fracture that divides the living rock. This is typically the result of contraction due to cooling or layering due to volcanic or other actions. These often run in parallel sets, and noting them carefully can help predict block sizing for extraction. A joint is never the result of shear displacement. In patternmaking, it is the point where stone units come together.

joint mold: The face template of a stone block (typically vertical) taken from the face that will abut the next stone in the course or arch.

jumper: The largest stone that occurs in random ashlar or random rubble patterns, used to "jump" the pattern up to the next length of horizontal coursework.

kerf: A slit made into a block or finished piece. Several slots or kerfs are sometimes made in series and then broken off as a quick way to roughly shape stone. Within the trade, it is often used incorrectly as a verb.

keystone: The central *voussoir* (wedge-shaped stone) of an arch or vault. This will sometimes extend below the *intrados* or forward beyond the plane of the arch. It is often carved or otherwise visually punctuated.

lacing course: The practice of turning the longest face perpendicular to the wall surface to bind the wall into the structural *hearting* behind; also called "bonding course." Individual stones used for this purpose are called "tie stones" or "dead-heads."

lime: A central ingredient in whitewash, plaster or render, and mortar, created by calcining limestone ($CaCO_3$) to a temperature of 900° C.

lime putty: A plastic and workable paste created by slaking lime (hydrating in water for three months to three years) that may be used as mortar or the base for *limewash*, plaster, or *render*.

limestone: A sedimentary stone composed of calcareous sediment, originating from the breaking up of shells, corals, and other marine animal life.

limewash: A coating for masonry walls made of lime and water. It is typically white in color, but sometimes tinted. It is generally applied over a *render* or plaster, coating the stone to create a smooth surface. Although generally waterproof, it is not particularly durable, and regular applications are to be expected.

lintel: The stone bridging a wall penetration such as a window or door. When spanning a fireplace opening, the lintel is often called the *mantel.*

liquid adhesion: In stone, water molecules' attraction to one another is weaker than their attraction to things they encounter. Thus, water will adhere to the underside of a stone coping or ceiling rather than drop off, unless a sufficient *drip kerf* or *throating* is added.

living rock: Undisturbed rock still connected to the earth, with groundwater (often called *quarry sap*) still migrating through it.

mantel: The spanning *lintel* over a fireplace opening.

mason's miter: A stone that has the appearance of a miter (a joint made by cutting two pieces of material at an angle and fitting them together), but instead is carved from a single stone. Mitered intersections are weak and liable to break, and were historically almost impossible to carve before industrial diamonds supplanted chisels in the 1950s. Avoiding miters extends the longevity of the stone as well as eliminating the *visual anachronism* resulting from their use.

mastaba: An ancient Egyptian stone tomb typology predating the pyramid, rectangular in form with sloping sides and a flat roof. Originally reserved for royalty, it was later adopted by well-to-do nobles and civic leaders.

menhir: A type of Neolithic vertical stone *monolith* erected primarily in northern Europe, usually without adornment.

metamorphic: Describing rock that has undergone transformation through intense heat, intense pressure, or folding; examples include hard limestone and marble. It is one of the three main geological categories of rock (the others are *igneous* and *aqueous*).

middle-third rule: The concept that the vertical or rising joint of the course below must land in the middle third of the current course being set. This simple method assures proper bonding or weaving of the pattern.

miter joint: The joint created when two oblique or beveled surfaces come together, meeting at an angle. The angle is often 90 degrees, but not always. Where the two pieces meet, the ends or bevels are equally angled so that the joint bisects the angle between the pieces. In stone masonry, miters are scarcely ever permitted between planes and never at *quoins.* See also *mason's miter* and *visual anachronism.*

modillion: A decorative stone "bracket" that adds support to a stone cornice.

Mohs hardness scale: A scale of mineral hardness based on which mineral will scratch another. The scale is ordinal, meaning that each number is twice as hard as the one prior. This is an ancient qualitative assessment mentioned as early as 300 BCE. The softest is talc (1), and the hardest is diamond (10).

molding: Typically carved or formed on the edge or face of a member to cast shadow or parallel stripes of light and shade. The surface may be sharply differentiated to project or recede and often uses simple or compound curves.

monolith: An exceptionally large block of stone. It may be worked minimally or formed into a monument such as an obelisk.

mortise: The carved hole (female element) of a two-part joint. See *tenon.*

non-extradosed: Arches whose *voussoirs* end on a curve and are not formally tied into the coursework containing them. See also *extradosed.*

obelisk: A monolithic, tall stone pillar with four tapered sides and a pyramidal cap. In modern times, the form has been commonly used to honor the dead.

orogeny: The primary geological mechanism driving mountain-building events. "Orogenesis" describes the full set of contributing processes.

parapet: Any section of wall occurring above the roofline, and typically composed of three elements: *plinth*, die or *dado*, and cap or *cornice.* It is also common on rampart walls, bridges, and balconies.

pedestal: A carved support for a statue, vase, or column, typically composed of three elements: *plinth*, die or *dado*, and cap or *cornice.* Not to be confused with the *base*, which is the lowest element of a column.

pediment: A plain, triangular space above the *architrave* in classical architecture. When carved, this is called a *frieze.*

pendentive: A concave triangular form that constitutes the transition and connection between a square form and a circular dome or polygonal shape.

perpend: A through-stone or bond stone that extends all the way through the wall, tying or locking its two faces together. In brickwork, the term applies to a brick turned on end, tying the exterior face to the wall's center (not required to reach the interior face, since the unit is often too short) and is sometimes called the *header*. The term is also used to describe the vertical mortar joint between two bricks.

petrichor: The scent of rain when it first falls on stone or dry soil.

pier: A strong square or rectangular support (usually of some thickness) supporting an arch, wall, or roof.

pilaster: A shallow imitation of a column that projects from a wall. It must be less than half of the column's normal thickness; otherwise, it is an *engaged column*.

pillar: A vertical support of any shape that is not identified as a column (by the classical orders).

pitching: The process of removing stone to flake the edge. When applied only to the edge of a slab of greater than one and one-quarter inch in thickness, the resulting split-faced texture is called simply a "pitched edge."

plane: In masonry, only the geometric definition applies: a flat or level surface, any section through which a like surface is a straight line.

plinth: An elevated stone pedestal or support, often used to lift columns or sculpture, raising and formalizing their placement. Also, the base of a building when treated differently (in terms of stone, pattern, or texture) from the walls above.

plug and feathers: A two-part tool consisting of two metal shims (feathers) that line a drilled or chiseled hole in a stone block. A wedge (plug) is set between the shims. When multiple holes and wedges are aligned with the grain of the stone and the wedges are tightened in series, the stone will split apart. Depending on resources and availability of metal, the feathers were often omitted (Italy, China, Japan), with the wedge leaving a particular telltale mark on the stone.

point chisel: The primary hand tool, shaped like a large, fat pencil, for rough-shaping a stone and removing excess material.

pointing: The process of packing a mortar joint between freshly laid stone or brick. It can also be used to describe the application of mortar to the surface of a joint to create an extruded bead. Such a bead guides the eye and suggests tight-joint mating of the individual stones, which may or may not actually exist. For important but subtle distinctions, see *re-pointing* and *tuck-pointing*.

polygonal work: A pattern assembled of irregularly shaped stones. These typically include combinations of irregular rectangles with five-sided and even six-sided shapes.

pozzolanic cement: In its natural form, a volcanic ash that when mixed with water will demonstrate cementitious bonding properties. The Minoan civilization pioneered its use, mixing it with calcined lime to create waterproof renders for baths, cisterns, and aqueducts, and the Romans used it extensively.

primary grain: Sometimes called the "natural bed" or *rift*, this is the grain direction along which a stone is easiest to split, since the majority of its particles are aligned along this orientation.

projecting boss, lifting boss: Small stone protrusion left for handling and/or the securing of ropes during transport or placement of large or difficult-to-maneuver shapes such as columns. Once the stones are in place, the projections in visible areas are carved down flush with the finished surface; finding them still intact offers an important clue to a structure's history, since it signals interruption by war or economic hardship.

proud: Describes a condition in which something lightly projects from a surface, as in *bas-relief*. Although uncommon outside stone and architecture, and distinctly British in origin, this term is widely used in the United States.

pseudisodomic: Describing alternating stone coursework of two differing heights, often seen in Greek and Roman ashlar. The smaller course is typically a bonding course that penetrates the structural *hearting* at the center of the wall. Some definitions require the individual stones of both courses to be of the same length.

pumice: A lightweight volcanic rock high in silica that is often used in concrete to create lightweight, nonstructural slabs or infill. It is also ground up and used as an abrasive for polishing.

quarry: An open excavation from which stone is removed. Used as a verb, it denotes the act of removing stone from its naturally occurring position.

quarry sap: The groundwater in stone, typically present during the first twelve months after a block is quarried.

quoin: Stones at the corners of a wall that have been sized larger to counteract the natural weakening inherent in turning a corner. Such stones will often receive a different surface treatment to highlight their scale or use, since they show a level of care and thoughtfulness typical of quality structures. They may be set flush with the wall, or set *proud* to highlight their formal shape. Antiquated and British spellings include "coin" and "coign."

raked joint: Common in rubble masonry, this is the result of clearing out or "striking back" the setting mortar prior to *pointing* the space between the stones with a high-quality finish mortar. Due to the weight of larger stones, the setting mortar is often made of coarse sand to add body while building the wall. The finish mortar will be of fine sand and may be colored to match the stone.

rampart: A strong wall surrounding a fortification, often including a *parapet* for defense.

random ashlar: Patterns of closely fitted ashlar blocks (joint widths of one-eighth inch or less) that do not run in continuous horizontal courses.

random rubble: Non–tightly wrought masonry that is not run in continuous horizontal courses. This pattern uses shapes that are typically more rectangular, differentiating it from polygonal work.

rebate: A longitudinal groove or channel carved into a stone, often at the edge, to define a consistent width where a door or window frame will fit. Sometimes called "rabbet." See also *draught*.

regolith: Unconsolidated rock that covers bedrock (as opposed to soil, which contains organic material).

relieving arch: An arch built above a *lintel* to help transfer the weight of the wall above to the lintel shoulders.

render: A coat of plaster or *limewash* serving a structural purpose, for instance, waterproofing and/or cooling. When nonstructural and decorative, the term "stucco" is used.

re-pointing: The process of placing fresh mortar into a scraped joint between older masonry (either stone or bricks), sometimes mistakenly referred to as tuck-pointing. For important distinctions, see *pointing* and *tuck-pointing*.

resinated slabs: Stone slabs that have been infused with epoxy resin because they are otherwise too weak to travel safely to market. The practice is also done to give the stone an appearance darker than nature provided it. Although this strengthens (and darkens) the stone in the short term, the resin will ultimately yellow after exposure and degradation by UV light. Even if the slabs are used indoors, UV light can penetrate windows, irreversibly discoloring the resin. Natural stone is colorfast.

retaining wall: A wall constructed to retain the earth, specifically designed to resist the lateral pressure of the material retained as well as the hydrostatic pressure of water moving through the earth. Retaining walls usually contain much more material (are thicker) than may be apparent from the wall face, and typically include significant drainage or "weep" holes to relieve the hydrostatic pressure that would otherwise accumulate.

rib: A narrow projecting member to strengthen a vault, ceiling, or panel.

rift: Of the three grain directions within stone (primary, secondary and tertiary), the easiest to split along, since the majority of particles from which the stone is composed are aligned within this orientation. Also called the *primary grain*.

riser: The vertical surface of a stair block. See also *tread*.

rising joint: The vertical joint between two stones of similar height.

rock: A solid mineral material component of the surface of the earth, technically in a state unaltered by human efforts (as opposed to *stone*).

rubble: Stone that is minimally shaped for building or walling. It can also denote a pattern style that is not tightly wrought. When used this way, it is typically accompanied by a pattern modifier: *snecked rubble*, random rubble, or random rubble built in courses.

running bond: The most basic pattern, for instance, in a typical brick wall, in which the vertical joints in the course below are covered by solid stones in the course above, so that the weaknesses created by the joints are capped by the spanning members of the second course.

rust jacking: The expansion of an iron support or wall tie due to rust, often with forces so powerful that concrete, ceramic, and many other human-made materials are blown apart. Most building stones are vulnerable as well.

rybat: The cut stone termination at the sides of windows and doorways used to resolve or terminate the larger stone pattern of the larger wall face; essentially a smaller version of a corner *quoin* at these smaller wall penetrations. It has been suggested that the term might also be a variation of "rabbet" (see *rebate*).

Sacred Geometries: The second and better-known rule set (after *Sacred Rules of Bondwork*) to come down to us from pharaonic ancient Egypt. It speaks to the universal math and geometry at the center of our natural and built environment. The premise rests on a belief that all form is determined by an invisible, immaterial world of pure geometry and ratio, linking such disparate examples as classical architecture to the golden section (one of the best examples of the Sacred Geometries) and the carbon bonding sequence.

Sacred Rules of Bondwork: One of two ancient rule sets originating in pharaonic ancient Egypt (see *Sacred Geometries*), through the cultures of ancient Greece and Rome, and finally codified by the Operative Guild of Freemasonry between 1050 and 1400 CE. The rule set enumerates the design principles considered core to the successful use of stone and was held as secret by initiated guild members.

sandstone: A sedimentary rock whose grains come from the degraded remains of other rocks (*aqueous*). One of the four primary building stones (the others are *granite* and *limestone*/marble, and slate).

sedimentary: Aqueous rock (such as *sandstone* and *limestone*) that traces its creation to a slow layering and buildup of mineral or organic sediment over time. These layers become fused through pressure and heat or are welded together by binding minerals carried by groundwater. This is one of the three main geological categories of rock (the others are *metamorphic* and *igneous*).

segmental arch: An arch whose curve of the *intrados* is less than 180 degrees (less than a half circle). It is generally held that a segmental arch must have a rise greater than one-eighth of the span to successfully resist the thrust.

semicircular arch: An arch whose curve of the *intrados* is a semicircle.

semi-elliptical arch: An arch whose curve of the *intrados* is half an ellipse.

shaft: The central element of a column, between the *base* and the *capital*.

sill: The bottom threshold of a door or window. These are typically sloped to shed water, extend beyond the wall, and contain a *drip channel* called a *throating*.

sneck, snecked rubble: A pattern of squares and rectangles that are not tightly fitted nor closely wrought, nor do they run in horizontal courses with a continuous joint. Furthermore, these typically contain "snecks" (small, usually square blocks) to help resolve the courses.

soffit: The horizontal, aloft underside of a construction element.

solid geometry: The branch of geometry used to illuminate three-dimensional objects.

span: The horizontal distance between the two sides of an arch or the unsupported length of a *lintel* or beam.

splay: A sloped surface, often making an intersection or oblique angle with another abutting angle, often at the sides of a door or window jamb.

splayed arch: An arched opening that tapers larger or smaller from the front to the back.

spolia: Historically, the product of the cultural theft of stones and stone objects by a conquering or colonial power. In modern usage it may be applied to the repurposing of antique stone or sculpture in new architectural applications.

springing: The point on the *intrados* of an arch or vault where the curve begins.

spring line: The line from which the first stones of an arch or vault rise.

stacked bond: When stone blocks are stacked directly on top of each other without the benefit of alternating the vertical joint from course to course, as in a *running bond*.

stele: A solid rectangular shaft or slab of stone with a rounded summit (not ending in a pyramidal top or obeliskoid form), often used as upright tablets to mark boundaries, milestones, or other commonplace uses.

stereotomy: The shaping of stone into complex geometric forms. The term derives from "solid-cutting."

stone: A *rock* that has been modified by carving, crushing, or other human exertions. Since stone is worked away from the compact lithic matter in the quarry, such a place is called a "stone quarry," whereas a "rock quarry" is a place where smaller unworked lithic material (such as glacial till) may be gathered.

stonecutting: The process of reducing stone to its desired shape and fit.

stretcher: A block of stone laid lengthwise, its longest face running horizontal to the face of the wall. If quarried and processed correctly, this face should show the primary grain structure of the block in section. The term is opposed to *header*.

string course: A continuous horizontal molding or projection around the perimeter of a building that may or may not be carved. Sometimes this projection is expanded and a *drip kerf* is added to the underside to help remove water from the facade face.

stylobate: A stepped platform supporting a row of stone columns.

swag: A festoon of fruit, flowers, or leaves that appears to hang on a facade, typically carved in *bas-relief* on a wall panel.

tenon: The projecting nub or tongue (male element) of a two-part joint. See also *mortise*.

theodolite: A modern surveying tool, informally called a "transit," constructed of a small telescope mounted on a leveling tripod for measuring vertical and horizontal angles.

throating: A carved *drip kerf* under a sill or coping.

through-stone: A individual stone that extends from one face of a wall to the other, through the *hearting* at the wall's center, effectively binding the two faces together.

tread: The horizontal surface of a stair block. See also *riser*.

trullo: A prime example of vernacular architecture, a small round residential building capped with conical roof. Although specific to the Itria Valley in the Puglia region of Italy, the construction technique can be traced back to Mycenae in the fifteenth century BCE. The plural is "trulli."

tuck-pointing: An advanced technique of placing two colors of mortar within a joint to give the illusion of finer workmanship. The main color matches the stone or brick, while the accent color conveys the illusion of a smaller joint. For important but subtle distinctions, see *pointing* and *re-pointing*.

tuff: A consolidated volcanic ash. The stone hardens over time, improving its structural integrity. *Tufo* in Italian.

ultimate strength: The maximum stress (often called the "tensile stress") that a material can endure before failing.

vault: A ceiling that is concave, often under an arch or series of arches. The term is also applied to underground chambers.

veneer: An outside, non-load-bearing *wythe* of masonry used as facing material.

viewing distance: A consideration that may determine the level of craftsmanship. Work only ever visible from thirty feet away may be rougher than work expected to be viewed at eye level and/or touched.

visual anachronism: The use or representation in one historical period of something that postdates the product's or technique's invention. For example, a mitered coping or wall panel in hard stone could not technically have been created prior to the introduction of industrial diamond saws in the 1950s. Although it was done occasionally in soft marble in the eighteenth century, it wasn't technically possible in harder materials. Therefore, to see such a technique used in a re-creation of English manor house architecture is visually jarring.

voussoir: A wedge-shaped, trapezoidal stone cut precisely and installed in series to form an arch.

weathering: An area sloped to drain water. Minimum slope for weathering is typically 5 percent.

weather joint: An external joint designed to keep out wind and rain. The joint is created by beveling or slanting downward and away, so that water will more readily shed off.

wythe: A vertical width of masonry one unit in thickness; in brickwork, a single thickness or course. A single wythe is not considered structural.

Notes

Introduction

1 Keith Critchlow, *Time Stands Still: A New Light on Megalith Science* (London: Gordon Fraser, 1979), 173.
2 Ami Ronnberg, Kathleen Martin et al., *The Book of Symbols: Reflections on Archetypal Images* (Cologne: Taschen, 2010), 104.
3 Annie Dillard ruminates, "We are civilized generation number 500 or so, counting from 10,000 years ago when we settled down. We are *Homo sapiens* generation number 7,500, counting from 150,000 years ago when our species presumably arose. And we are human generation number 125,000, counting from the earliest *Homo* species." Annie Dillard, *For the Time Being* (New York: Knopf, 1999), 187.
4 The modern Masonic guilds differentiate between the "operative" branch of freemasonry (those that work with the tools) and the "speculative" branch. Although they arose out of a single tradition, they diverged as more speculative Freemasons joined the guild in the early seventeenth century. Douglas Knoop writes of Scottish lodge records of the seventeenth century containing numerous examples of "non-operative members." Douglas Knoop, *Pure Antient Masonry* (Manchester, UK: Manchester University Press, 1939), 38. See also Robert Freke Gould, *The History of Freemasonry*, vol. 1 (New York: John C. Yorston, 1886). For the early history of the guild see W. Ravenscroft, *The Comancines* (London: Elliot Stock, 1910).
5 In the EU, the weight of standard cement bags has now been reduced to 25 kilograms to reduce worker injuries.
6 The ancient chestnut beams supporting the stone floors of medieval buildings in Siena cannot support the weight of a modern hotel bathroom. Most of the real work of this initial project was to install steel wide-flange beams across the ceilings of the interior spaces. Supported by the thick outside walls of the building, the suite's individual bathrooms effectively hung from these spanning steel members, their weight thus transferred to the bearing walls of the structure.
7 Eventually I was able to transfer the sand and bags in just over ten hours (twelve minutes per trip), making my pay the equivalent of just under two US dollars per hour, or almost twenty dollars per day.
8 Oliver Faude, Wilfried Kindermann, and Tim Meyer, "Lactate Threshold Concepts; How Valid Are They?," *Sports Medicine* 39, no. 6 (2009): 469–90. The lactic threshold (lactate inflection point, or LIP) is the exercise intensity point at which your bloodstream starts to rapidly produce lactic acid. Working below the lactate threshold allows the body to process the acid and avoid its buildup in the muscles. It is estimated that working for one hour at one's threshold (say, 148 heartbeats per minute) is the equivalent of working for two minutes above the threshold (for instance, 160 heartbeats per minute) or for four hours under the threshold (140 heartbeats per minute).
9 Those who have acquired their knowledge outside formal channels of education are sometimes called "reflective practitioners."
10 The *Campinion* of France offer the last true apprenticeships, with ties to the ancient guilds of medieval times.
11 "The Ancient geometries were protected by a pledge of utter secrecy, passed from the temples to various guilds and craftsmen." Tons Brunés, *The Secrets of Ancient Geometry and Its Use* (Copenhagen: International Science Publishers, 1967), 1:12. See also Nicola Coldstream, *Medieval Craftsmen: Masons and Sculptors* (1998; repr., Toronto: University of Toronto Press, 1991), 12–13; Francis B. Andrews, *Further Notes on the Medieval Builder* (1925; repr., New York: Barnes and Noble, 1993), 6.
12 One of the primary surviving guild documents, known as the Regensburg Ordinances, was the result of a conclave of master masons from around Germany who gathered in Regensburg in 1459. The Ordinances are explicit in their expectation of guild secrecy. Gould, *Hist. of Freemas.*, 146–47. See also Coldstream, *Med. Craftsmen*, 10–13, 21, 38.
13 I. E. S. Edwards, *The Pyramids of Egypt* (1961; repr., New York: Viking, 1972), 202. See also Somers Clarke and R. Engelbach, *Ancient Egyptian Masonry: The Building Craft* (New York: Dover Publications, Inc., 1930; repr., 2010), 25.
14 Denys A. Stocks, *Experiments in Egyptian Archaeology: Stoneworking Technology in Ancient Egypt* (New York: Routledge, 2003), 60. Stocks indicates that only cold annealing prevented damaging cracks from appearing.
15 "The existence of trade secrets that were passed on encouraged a family adherence to the calling of architecture. Sons of architects learned the recondite language from their fathers and taught it to their own sons. A professional dynasty, if not perhaps strictly lineal, could thus be traced among practitioners of architecture, much like the recorded order of royal dynasties." Spiro Kostof, "The Practice of Architecture in the Ancient World: Egypt and Greece," in *The Architect: Chapters in the History of the Profession*, ed. Spiro Kostof (New York: Oxford University Press, 1977), 6. A similar parallel can be traced in the handing down of "mason's marks," unique symbols of ownership carved into wrought stone pieces by individual masons. John Harvey notes that one of the elder masons at Canterbury Cathedral reported in 1844, "His own mark belonged to his father and grandfather before him, and this 'his grandfather had it from the lodge.'" John Harvey, *The Master Builders: Architecture in the Middle Ages* (New York: McGraw Hill, 1971), 47. See also Coldstream, *Medieval Craftsmen*, 13.
16 William Bryant Logan tells the story of a thousand-year-old Scottish cemetery where the marble gravestones lost 3.5 inches in height. The granite monuments "lost only one tenth that amount." William Bryant Logan, *Dirt: The Ecstatic Skin of the Earth* (New York: Riverhead Books, 1995), 121.
17 In Texas, state agencies are required to "componentize [*sic*] various elements of buildings in order to calculate and record depreciation of a building's structural components, subsystems and equipment." This requirement was passed in 2001 (SB 482) by the state legislature. The goal of the statute is to help officials evaluate whether project expenditures would be cost-effective as related to the "useful life" of the building, which is typically estimated to be twenty years. Francis Duffy, "Measuring Building Performance," *Facilities* 8, no. 5 (May 1, 1990): 17–20, https://doi.org/10.1108/EUM0000000002112.

Chapter One

1 Harvey, *The Master Builders*, 9.
2 Peter Rockwell, *The Art of Stoneworking: A Reference Guide* (Cambridge, UK: Cambridge University Press, 1995), 9.
3 Nicholas Wade, "Those Ancient Incan Knots? Tax Accounting, Researchers Suggest," *New York Times*, August 16, 2005, https://www.nytimes.com/2005/08/16/science/those-ancient-incan-knots-tax-accounting-researchers-suggest.html. See also Vincent R. Lee, *Design by Numbers: Architectural Order among the Incas* (Cortez, CO: Sixpac Manco, 1996).
4 Stella Nair, US archaeologist, personal correspondence, October 2020. Nair observes that although the Inca produced some bronze for weapons, their culture is "better described as Stone Age."
5 Visiting Cuzco in 2005, I was still able to purchase several meteorite hammerstones from local shops who offered them for sale. The sellers seemed baffled by them and suggested they were thrown from ancient slings. Although they range in size, they are typically the size of a small fist, with the top worn perfectly smooth from generations of hard use.
6 Jean-Pierre Protzen, "Inca Quarrying and Stonecutting," *Journal of the Society of Architectural Historians* 44, no. 2 (1983): 165.
7 Protzen, "Inca Quarrying and Stonecutting," 188.
8 Jean-Pierre Protzen and Stella Nair, *The Stones of Tiahuanaco: A Study of Architecture and Construction* (repr., 2014, Los Angeles: Cotsen Institute of Archaeology Press, 2013), x. Protzen cites the Early Intermediate and Late Intermediate Periods.

9 Protzen and Nair, *The Stones of Tiahuanaco*, xii.
10 Henry Hodges, *Technology in the Ancient World* (New York: Alfred A. Knopf, 1970), 92.
11 Rockwell, *The Art of Stoneworking*, 199.
12 Edwards cites a discovery by W. B. Emery in the Egyptian cemetery of Saqqara, where, in the First Dynasty, wet abrasive material such as moistened quartz sand, which is plentiful in Egypt, may have been employed. Edwards, *The Pyramids of Egypt*, 201. Norman Davey further illuminates the method: "There is evidence from the markings left on granite statuary that cutting and drilling were possible. Cutting may have been achieved by using an abrasive such as emery, fed to a soft metal blade, and drilling by feeding the abrasive down a soft copper tube, a technique that has survived to this day for boring holes in glass. The hard grains of abrasive bedded themselves in the soft copper." Norman Davey, *A History of Building Materials* (London: Phoenix House, 1961), 15.
13 Paul N. Hasluck, *Building Stones* (London: Cassell, 1904), 28. Hasluck adds that once "chilled iron" or shot was substituted for sand, productivity soared from three-quarters of an inch every two days to one and a half inches per hour (citing the cutting of a four-foot-long granite block) (88).
14 Tullia Ritti, Klaus Grewe, and Paul Kessener, "A Relief of a Water-Powered Stone Saw Mill on a Sarcophagus at Hierapolis and Its Implications," *Journal of Roman Archaeology* 20 (2007): 138–63, esp. 148, fig. 10.
15 Davey, *A History of Building Materials*, 15.
16 As the extraordinary ninth-century Irish chapel built on Saint Macdara's Island demonstrates, one can always find exceptions. Be that as it may, it remains useful to draw general conclusions from relevant historical time periods.
17 Jean Gimpel, *The Cathedral Builders* (New York: Grove Press, 1961), 5.
18 Gimpel, *The Cathedral Builders*, 6.
19 Alain Erlande-Brandenburg, *The Cathedral Builders of the Middle Ages* (London: Thames & Hudson, 1995), 15–16.
20 G. G. Coulton, *Art and the Reformation: Part I, Medieval Faith and Symbolism* (New York: Harper Torchbooks, 1928; repr., 1958), 100–101.
21 Coulton points out that in the Gothic period, "ornamentation [was] mainly structural." He further claims, "Substantial unanimity among writers who differ widely on other points; Gothic art began to decline as soon as ornamentation became superficial rather than structural; rather a veneer than an essential constituent of the building itself." Coulton, *Medieval Faith and Symbolism*, 11–12.
22 Erwin Panofsky, *Abbot Suger on the Abbey Church of St.-Denis and Its Art Treasures* (Princeton, NJ: Princeton University Press, 1948), 59.
23 Panofsky, *Abbot Suger on the Abbey Church of St.-Denis and Its Art Treasures*, 51. For an inspired interpretation of the Latin poetry of this inscription as informed by the phraseology of John the Scot, see pages 22–23 in the same text.
24 Gimpel, *The Cathedral Builders*, 5.
25 Charles George Herbermann, "Basilica of St. Peter," in *The Catholic Encyclopedia*, ed. Charles George Herbermann (New York: Encyclopedia Press, 1907; repr., 1913), XIII:369.
26 Vincent R. Lee, *Ancient Moonshots: Megalithic Mysteries from before Technology* (Jackson Hole, WY: Sixpac Manco, 2013), 166–77.
27 Alexis A. Julien reports that it was found broken in two and "re-cemented." Although I assume this happened after its reinstallation in Rome, the author does not state when this break and repair took place. Alexis A. Julien, "The Misfortunes of an Obelisk," *Journal of the American Geographical Society of New York* 25, no. 1 (1893): 91.
28 Curran et al., *Obelisk*, 54. Constantine moved the obelisk to Alexandria, but died in 337 CE before it could be brought to Rome. His son, Emperor Constantius, completed the transfer and erection more than two decades later. Also, although obelisks capture the world's attention, the longest piece of stone to have been successfully quarried and erected was put up by the subjects of the Kingdom of Aksum, an ancient Ethiopian civilization. Known as the Great Stele at Axum, now fallen, it measures 108 feet tall and is estimated to weigh 520 tons. This puts the Lateran Obelisk at a close second in length and weight. Steles differ from obelisks in that steles don't conclude with a pyramid at the top.
29 Brian A. Curran et al., *Obelisk: A History* (Cambridge, MA: MIT Press, 2009), 55.
30 Fontana Domenico, *Della Trasportatione dell'Obelisco Vaticano* (Rome: 1590; repr., Oakland, CA: Octavo, 2002), Rowland commentary, 4.
31 Henry J. Cowan, *The Master Builders: A History of Structural and Environmental Design from Ancient Egypt to the Nineteenth Century* (New York: John Wiley & Sons, 1977), 189–92.
32 Domenico, *Obelisco*, Rowland commentary, 12.
33 Sigfried Giedion, *The Eternal Present: The Beginnings of Architecture* (New York: Pantheon Books, 1964), 2:440.
34 Robert Lawlor, *Sacred Geometry: Philosophy and Practice* (London: Thames & Hudson, 1983), 30.
35 Radiocarbon dating has indicated much earlier habitation of Easter Island. The dates cited reflect the general consensus on our current (imperfect) understanding. See John Flenley and Paul Bahn, *The Enigmas of Easter Island: Island on Edge* (New York: Oxford University Press, 1992), 77.
36 The largest moai ever completed and installed is thirty-two feet tall and is called Paro by the local population. Katherine Routledge, *The Mystery of Easter Island* (1919; repr., New York: Cosimo Classics, 2007), 113. For a fascinating account of the how the moai were likely moved, see Lee, *Ancient Moonshots*, 101–38.
37 Eugène Emmanuel Viollet-le-Duc, *Discourses on Architecture*, trans. Benjamin Bucknall (1889; repr., New York: Grove Press, 1959), 1:294.
38 Cowan explains that despite this reliance on the scale and mass of the stone, the strength of the material was rarely fully utilized. Cowan, *The Master Builders*, 36.
39 Brad Goldberg, US stone sculptor, personal conversation, February 2008.
40 "Blocks and quoins [corner stones] were supplied ready cut to many sites, including Caernarfon Castle, and quarry identification marks have been found on blocks from the big French quarries such as Caen and Tonnerre." Coldstream, *Medieval Craftsmen*, 40.
41 Data from my manufacturing operations have shown that the loss of material from block to finished product can approach 65 percent depending on the tolerances and pattern. Interestingly, when the great archaeologist William Matthew Flinders Petrie was surveying at Giza, he found chips of stone discarded by the masons and dumped over the sides of the cliffs to the north and south of the Great Pyramid. He described them as "probably equal in bulk to more than half of the Pyramid." That, of course, included only the waste that was created at the site and not during the pre-shaping of the material at the quarry. Edwards, *The Pyramids of Egypt*, 217.
42 William Bryant Logan, *Dirt: The Ecstatic Skin of the Earth* (New York: Riverhead, 1995), 33–34, 91.
43 Edwards, *The Pyramids of Egypt*, 39.
44 Edwards, *The Pyramids of Egypt*, 51.
45 Harold North Fowler, James Rignall Wheeler, and Gorham Phillips Stevens, *A Handbook of Greek Archaeology* (New York: American Book Company, 1909), 117–18.
46 William Bell Dinsmoor, *The Architecture of Ancient Greece: An Account of Its Historic Development* (1909; repr., London: B. T. Batsford, 1950), 176.
47 Viollet-le-Duc, *Discourses on Architecture*, 1:287.
48 Rowland J. Mainstone adds, "The earliest columns were, almost certainly, saplings, the branches of larger trees, or bundles of reeds, simply thrust into the ground like the stakes of a modern fence to give them some resistance to overturning." Rowland J. Mainstone, *Developments in Structural Form* (Cambridge, MA: MIT Press, 1975), 165.
49 Edwards, *The Pyramids of Egypt*, 46.
50 I am thinking here of Selim Hassan's excavations in 1932–33. See Edwards, *The Pyramids of Egypt*, 123.

51 Julien, "The Misfortunes of an Obelisk," 94.
52 Reginald Engelbach, *The Problem of the Obelisks From a Study of the Unfinished Obelisk at Aswan* (New York: George H. Doran, 1923), 85.
53 Engelbach, *The Problem of the Obelisks*, 85.
54 Julien, "The Misfortunes of an Obelisk," 92–93. Engelbach repeats the story, and adds: "If this Rhamsesis was Ramesses II, the loss of a son would not have been vital, as he is known to have had over a hundred, to say nothing of several score daughters!" Engelbach, *The Problem of the Obelisks*, 91.
55 Tom Williamson, *Inigo's Stones: Inigo Jones, Royal Marbles and Imperial Power* (Padstow, Cornwall, GB: Matador, 2012), 58, citing Italian mineralogist Francesco Rodolico.
56 Julien, "The Misfortunes of an Obelisk," 91–92. Many scholars suggest it is water rather than oil being poured in front of the sledge, but to anyone with experience, this theory makes little sense.
57 Vincent R. Lee, US architect and archaeologist, personal conversation, November 2004.
58 Malcolm Hislop, *Medieval Masons* (Oxford: Shire, 2009), 31.
59 "Ultimately drills were available that turned the bit slightly with each stroke to ensure that the bit struck fresh rock, and that advanced themselves automatically as the rock was cut away. By the early twentieth century, drills could average 1.75 inches per minute, or 105 inches an hour, six-and-a-half times the rate for hand drilling." Thomas F. Mahlstedt and Lauren Cook, *Granite Railway Quarry, Phase II Stages 1 & 2* (Boston: Metropolitan District Commission, 2003), 3:37.
60 Paul Wiseman, US interior designer, personal conversation, March 2002.
61 Erwin Panofsky, *Gothic Architecture and Scholasticism* (New York: Meridian Books, 1959), 38.
62 Whether this makes any rational sense from a design perspective is a question I will address forthwith.
63 Aon Center in Chicago was built to house the headquarters of Standard Oil in 1974. In an excellent example of the iterative quality of masonry knowledge, the forty-three thousand original panels of Carrara marble that originally clad the building were removed due to warping and attachment failure (the warping of true marbles is called "hysteresis"). These were replaced with two-inch-thick granite from the Mount Airy quarry at a cost of approximately $80 million between 1990 and 1992. The retired Carrara panels were crushed and used as decorative gravel around the Amoco Refinery in Whiting, Indiana.
64 Data for the year 1999 showed 1.25 million tons. US Geological Survey, *Mineral Commodity Summaries* (Reston, VA: US Geological Survey, 2019), 156.
65 Edwards, *The Pyramids of Egypt*, 138.
66 "City Hall: The Building," http://www.aviewoncities.com/philadelphia/cityhall.htm. Wall thickness for stone walls previously considered structural are more typically fifteen to twenty-one inches. In many countries, twenty-one inches is considered standard thickness. Thinner than fifteen is unusual, since the wall is not easily bonded together to provide adequate strength. Further, Mainstone adds, "Quite recently it was still the almost universal practice to give a wall a thickness that was not less than, say, 1/16th of its height." Mainstone, *Developments in Structural Form*, 170.
67 Surface area to tonnage is an interesting metric and offers insight into how technology has allowed successive civilizations to radically expand the use of stone over the millennia. For example, each face of the Great Pyramid at Giza exceeds five acres, or almost 220,000 square feet (868,000 square feet total); a face, in this instance, would be one of the sloped pyramidal sides. When we divide that square footage by the weight of the pyramid (estimated at 2.3 million metric tons), we arrive at roughly 2.7 metric tons (6,000 pounds) per square foot of finished face. Contrast that with the average eighteen-inch thickness of stone walls during the 1920s, averaging 260 pounds per square foot of surface face. Today, less than a hundred years later, the weight per square foot of surface face has dropped to an astonishing 18 pounds per square foot. In practical terms, the stone that is quarried today will cover almost 11.5 times more surface area than one hundred years earlier, and 333 times more than during the ancient Egyptian civilization.
68 Seattle's King Street Train Station has a glorious masonry tower and well-proportioned interior moldings that for almost sixty years were obscured by a modern "drop" ceiling. Happily, the station was restored in 2013.
69 The earthquake of 2016 measured 6.2 on the Richter scale and caused 299 deaths. The earthquake of 2008 in L'Aquila in the Abruzzo region killed more than 300 and displaced about 65,000 people.
70 Panofsky, *Gothic Architecture and Scholasticism*, 8.
71 Estimates vary between 13,000 and 20,000 citizens lost, and approximately 400,000 structures destroyed. US Geological Survey, *Preliminary Earthquake Report*, archived from the original on November 20, 2007.
72 Edwards, *The Pyramids of Egypt*, 108.
73 Viollet-le-Duc, *Discourses on Architecture*, 1:375.
74 Edward R. Ford, *The Details of Modern Architecture* (Cambridge, MA: MIT Press, 2003), 1:3.
75 E. G. Warland, *Modern Practical Masonry* (London: Library Press Limited, 1929), 16.

Chapter Two

1 Donald Friedman, US stone engineer, personal conversation, May 2007.
2 Brunés, *The Secrets of Ancient Geometry and Its Use*, 1:9, 12.
3 A. Giry, *Notice sur un traité du Moyen âge, intitulé: De Colorisbus et Artibus Romanorum*, 1878, Bibliothèque de l'École des Hautes Études, fasc. 35, Paris; John Chatterton Richards, "A New Manuscript of Heraclius," *Speculum* 15, no. 3 (July, 1940): 225–71.
4 Coulton, *Medieval Faith and Symbolism*, 95–96. For the most recent research on the date of Eraclius see Mary P. Merrifield, ed., *Original Treatises on the Arts of Painting* (1849; repr., New York: Dover, 1967), 1:vii.
5 The documentary evidence for these conclaves starts in 1459 in Regensburg; in 1462 they continued in Torgau, and then alternated between Speyer and Strasbourg in 1464, 1466, 1467, 1468, and 1471. These are often collectively referred to as the Regensburg Statutes. Coulton considers the earlier statutes to have been refined and finally ratified in 1563 at Strasbourg by the largest gathering ever recorded: seventy-two masters and thirty journeymen. Coulton, *Medieval Faith and Symbolism*, 126. Beyond the Regensburg Statutes, significant written records occurred outside Germany in York, England, in 926, Milan in 1398, and Canterbury, England, in 1429. These gatherings "united a smaller or larger number of experienced men who exchanged their knowledge and discussed the standards of the art." Gerhard Rosenberg, "The Functional Aspect of the Gothic Style, Part II," *Journal of the Royal Institute of British Architects*, no. 43 (February 8, 1936): 364–71, esp. 365.
6 Article LIV, "Ordinances and Articles of the Fraternity of Stonemasons." This became the most widely distributed collection of the German stonemasons' ordinances. Gould, *The History of Freemasonry*, 1:129.
7 Brunés, *The Secrets of Ancient Geometry and Its Use*, 1:10.
8 Francis B. Andrews, *The Mediaeval Builder and His Methods* (1925; repr., New York: Barnes and Noble, 1993), 10.
9 Much of what we know of the oldest guilds of the Freemasons comes from manuscripts of rules and regulations dating from the fourteenth and fifteenth centuries. English statutes known as the Articles and Points of Masonry are thought to have been composed around 1430 but are generally regarded as being a copy of a document a century older, which in its turn is regarded as being based on even older statutes. See Jan Svanberg, *Master Masons* (Sweden: Carmina, Ltd., 1983), 84. The Articles and Points of Masonry state explicitly that the rise of masonry was due to "Great lords' children freely begotten" and they forbade a master to take any apprentice born of bond blood. Harvey, *The Master*

Builders, 48. The German statutes go much further: "The first condition, preliminary to binding an apprentice, was that he should prove his legitimate birth (art. LX.)… that his progenitors had been freemen for at least two generations, and that they had not followed any trade which was, in the eyes of this particular trade, degrading." Gould, *The History of Freemasonry*, 1:145.

10 Apprenticeships were traditionally seven years in length. They were shortened to six, and then five, over the last two hundred years. Coulton suggests that the *journée* mainly worked for day wages, and the journeyman title was the equivalent of a university bachelor's degree. The master was distinguished by his cap and gown, as seen on the tombstone of Hugues Libergier (d. 1263 CE) at Reims, France. Coulton, *Medieval Faith and Symbolism*, 139.

11 Andrews notes, "One of the earliest uses of the word 'master' occurs in 1127 when one Andrew was *cementarius* at St. Paul's and was doubtless a 'master.' The next mention is of *Magister Robertus, cementarius* in connexion with Westminster, and afterwards the term is in frequent use." Andrews, *The Mediaeval Builder and His Methods*, 53.

12 Andrews, *The Mediaeval Builder and His Methods*, 10; see also Coldstream, *Medieval Craftsmen*, 15; Svanberg, *Master Masons*, 89.

13 Gould, *The History of Freemasonry*, 1:144.

14 Svanberg writes, "Not until this stage did the mason learn the finer arts of masonry, such as carving out foliage and figures from the stone (*umb Kunst dienen, als auszugen, Steinwerg, Laubwer, oder Bildnuss*, as the German statute expresses it)." Svanberg, *Master Masons*, 89–91.

15 The previously cited Regensburg Statutes detail the professional advancement from apprentice to journeyman to master mason and offer insight into other aspects of professional conduct. Coldstream, *Medieval Craftsmen*, 10–13, 21, 38.

16 Harvey, *The Master Builders*, 30–31.

17 In a footnote, Coulton, *Medieval Faith and Symbolism*, 173, cites Ferdinand Janner, *Die Bauhütten des deutschen Mittelalters* (Leipzig: E. A. Seemann, 1876), who reports the entrepreneurial activities of the master mason of the cathedral at Regensburg, Conrad Roritzer (1450 CE), who was working multiple jobs and doing commissions and personal piecework on the side. For insight on social class, see Svanberg's discussion on how the "perspective of rank makes a clear distinction between the great patron and the small artist." Svanberg, *Master Masons*, 33.

18 Svanberg, *Master Masons*, 100.

19 The legal right to "jurisdiction over itself" was conferred first by the German King Rudolf I of Hapsburg in 1275 and later extended to the English lodges for "self-regulation." Svanberg, *Master Masons*, 86–87.

20 Rosenberg points out that the "secrecy was not universally maintained in the later period as can be seen from the translation of Vitruvius by Cesariano 1521, where he inserted rules for the setting out of the cathedral at Milan." Gerhard Rosenberg, "The Functional Aspect of the Gothic Style, Part I," *Journal of the Royal Institute of British Architects*, no. 43 (1936): 274.

21 R. A. Schwaller de Lubicz and Lucie Lamy, *The Temple in Man: The Secrets of Ancient Egypt* (Brookline, MA: Autumn Press, 1946; repr., 1977), 106.

22 The works published by German master masons include Mathes Roriczer's *Buchlein von der Fialen Gerechtigkeit* (Booklet on the Correct Design of Pinnacles, 1486) and *Geometria Deutsch* (1487); Hans Schmuttermayer's *Fialenbuchlein* (Schmuttermayer was a contemporary of Roriczer, though the date of his publication is uncertain), and Lorenz Lechler's *Underweisung* (Instruction, 1516).

23 Harriet Crawford, *Sumer and the Sumerians* (1991; repr., Cambridge, UK: Cambridge University Press, 2000), 78; C. E. Larsen, "The Mesopotamian Delta Region: A Reconstruction of Lees and Falcon," *Journal of the American Oriental Society* 95 (1975): 43–57.

24 In 1824, Joseph Aspdin invented Portland cement by burning ground soft sulfate mineral (gypsum) with finely divided clay in a lime kiln until the carbon dioxide was "driven off." It was then ground up. The name comes from the English village of Portland, known for its fine gray cement-colored building stone. Portland cement was originally a marketing name to imply that the result was as handsome as the stone from Portland.

25 Merrifield, *Original Treatises on the Arts of Painting*, 1:210. Merrifield acknowledges that the story is widely reported by multiple sources, but voices reservations about its veracity.

26 Christopher Payne, "Vital Vessels," *New Yorker*, December 7, 2020, 48.

27 Cowan, *The Master Builders*, 122; Stefano Carboni, personal conversation, 2015.

28 James W. P. Campbell, *Brick: A World History* (London: Thames & Hudson, 2004), 303.

29 Cowan, *The Master Builders*, 204. See also Rosenberg, "The Functional Aspect of the Gothic Style, Part II," which claims that structural mechanics did not exist as a science until 1600 and that "the proportions of structural members were expressed by formulae which were made from experience and laid down…in figures" (366).

30 Coulton, *Medieval Faith and Symbolism*, 7.

31 Svanberg, *Master Masons*, 30.

32 Coldstream, *Medieval Craftsmen*, 51; Svanberg, *Master Masons*, 28.

33 Coulton, *Medieval Faith and Symbolism*, 89.

34 A modern structural analysis of Beauvais Cathedral suggests that its collapse was due to the lack of sequencing the loads on the structure during its construction. The loading of masonry structures is known as "equilibrium" in the Sacred Rules of Bondwork. Also see Jacques Heyman, "Beauvais Cathedral," *Transactions of the Newcomen Society* 40, no. 1 (1967–68): 15–35. "Thus the fabric of Beauvais in 1272 seems to have been, in the large, designed almost perfectly to fulfill its function.…The possibility therefore cannot be overlooked that the collapse began with some trivial accident and spread thence to the whole of the fabric" (30). Over the years, many have speculated that the sheer size of the edifice led to its demise. However, Heyman continues, "[Robert] Branner has pointed out that, despite all that has been written about the colossal dimensions of Beauvais, they are not much greater in fact than those of the great cathedrals of the first half of the thirteenth century" (18).

35 Originally constructed in 1176–1209, the bridge finally failed entirely in 1820. Cowan, *The Master Builders*, 95.

36 Sir Banister Fletcher, *History of Architecture*, 19th ed. (London: Butterworths, 1896; repr., 1987), 286.

37 "[In] 1282 labour was impressed on a wholly new scale: 1000 diggers, 345 carpenters and 50 masons gathered at Chester and Bristol, and were sent into Wales to start preliminary work on the castle sites." Coldstream, *Medieval Craftsmen*, 20.

38 Cowan, *The Master Builders*, 96.

39 "Report Cites Design and Structural Faults as Cause of Paris Airport Collapse," *Insurance Journal*, February 16, 2005; "Paris Air Terminal Collapse Report," *Architecture Week*, April 27, 2005, N1.2.

40 Coulton, *Medieval Faith and Symbolism*, 74. It may prove informative to contrast such religious superstition with the cool rationalism of the twentieth century. As late as 1966, the following theory was put forth by civil engineer Jacques Heyman: "If on striking the centering for a flying buttress, that buttress stands for five minutes, then it will stand for 500 years." Jacques Heyman, "The Stone Skeleton," *International Journal of Solids and Structures* 2, no. 2 (1966): 249–79, esp. 254. The term "centering" refers to the wood form supporting the stonework during its assembly.

41 David Herbert Somerset Cranage, *Cathedrals and How They Were Built* (Cambridge, UK: Cambridge University Press, 1951), 8.

42 Amy Gansell, personal correspondence, 2005.

43 Cowan, *The Master Builders*, 27.

44 Joseph W. Lstiburek, "Mind the Gap, Eh?," *ASHRAE Journal* 52, no. 1 (January 2010): 57–63, available at https://buildingscience.com/documents/insights/bsi-038-mind-the-gap-eh.

45 Heyman, "Beauvais Cathedral," 15.

46 Schwaller de Lubicz and Lamy, *The Temple in Man*, 106.

47 To which the Italians, having centuries of practiced nonchalance, replied, "Science without art is nothing."
48 See James S. Ackerman, "*Ars sine scientia nihil est*, Gothic Theory of Architecture at the Cathedral of Milan," *Art Bulletin* 31 (1949): 84–111.
49 Les Parrott III, personal conversation, 2004.
50 Ben R. Rich and Leo Janos, *Skunk Works* (New York: Little, Brown, 1994), 29. This concept of "visual safety" was also a maxim for aircraft designer Kelly Johnson: "An airplane that looked beautiful would fly the same way."
51 The archaeologist Adriana von Hagen, reflecting on her experience with ancient Inca architecture, points out that economic and personal insecurity are common denominators of human experience. As such, they also operate as chief drivers for choosing where we locate our structures and the quality of the building materials we use to construct them. Adriana von Hagen and Craig Morris, *The Cities of the Ancient Andes* (London: Thames & Hudson, 1998), 34.
52 Brunés, *The Secrets of Ancient Geometry and Its Use*, 1:11–12.
53 Cowan, *The Master Builders*, 34.
54 William Richard Lethaby points out, "Early inventions must have seemed like revelations, and skilled craftsmen were looked upon as magicians." William Richard Lethaby, *Architecture: An Introduction to the History and Theory of the Art of Building* (1911; repr., London: Thornton Butterworth, 1929), 19.
55 The original definition provided by Dinsmoor is particularly useful: "The entasis had the purpose of correcting a disagreeable optical illusion, which is found to give an attenuated appearance to columns formed with straight sides, and to cause their outlines to seem concave instead of straight." Dinsmoor, *The Architecture of Ancient Greece*, 169. The Greeks used entasis widely, although the starting point of fourteen feet in height for a column requiring entasis was somewhat variable. See Fowler, *A Handbook of Greek Archaeology*, 113n2. Also, the bowing of a column in the upper two-thirds is different than the gradual tapering of a column, which is also common but not defined as entasis.
56 "One difficulty in interpreting the architecture of the past is the deplorable habit of many conquerors of demolishing prestigious buildings and raising their own most important buildings on the ruins; this not merely destroyed the building but made excavation of the remains more difficult. Presumably this was done partly as a mark of triumph and partly to attract the loyalty attaching to the site. In recent years the foundations of the Great Pyramid, the most important Aztec religious monument, were found under the cathedral in Mexico City. Justinian's Palace has been excavated under the Blue Mosque in Constantinople, and the remains of a Roman temple are presently being excavated under the Duomo of Florence." Cowan, *The Master Builders*, 7.
57 Lethaby, *Architecture*, 206.

Chapter Three

1 Alfonso Acocella, *Stone Architecture: Ancient and Modern Construction Skills* (Lucca, Italy: Lucense SCpA, 2006), 210.
2 The time it takes for stone to cool is a complex question dependent on many factors, including the depth of its location (at twelve miles underground, it may take millions of years) and its relation to conductive cooling and associated permeability. See Andrew A. Snelling and John Woodmorappe, "The Cooling of Thick Igneous Bodies on a Young Earth," in *Proceedings of the Fourth International Conference on Creationism*, vol. 4, article 46 (1998), 527–45, https://digitalcommons.cedarville.edu/icc_proceedings/vol4/iss1/46.
3 Archibald Geikie, *Text-Book of Geology* (1882; repr., New York: MacMillan, 1903), 663. George A. Mitchell and A. M. Mitchell use the language "planes of stratification, called in this case planes of cleavage," in *Building Construction, Part 2, Advanced Course*, 16th ed. (London: B. T. Batsford, 1947), 73.
4 John McPhee, *Basin and Range* (New York: Farrar, Straus and Giroux, 1981), 7.
5 T. Nelson Dale, *The Chief Commercial Granites of Massachusetts, New Hampshire and Rhode Island* (Washington, DC: Government Printing Office, 1908), 19.
6 Dale, *The Chief Commercial Granites of Massachusetts, New Hampshire and Rhode Island*, 19.
7 Fred Whitridge, personal correspondence, December 2020. Whitridge, of the Evergreen Slate Company, reports that most of their Vermont slate is quarried at 45 degrees.
8 Lee, *Ancient Moonshots*, 102–6.
9 Mainstone also uses this term while speaking of the Egyptian pyramids: "Such heterogeneous masses have natural angles of repose related to the shapes of the individual blocks." Mainstone, *Developments in Structural Form*, 188–89. See also McPhee, *Basin and Range*, 26.
10 W. Lewis Barlow, "Historic Perspectives III: The Role of Transportation and Technology in the Development of the New England Stone Industry" (lecture presented at "Building with Stone: Granite and Marble for Architectural Exteriors and Monuments," MIT, Cambridge, MA, 2004). The story is corroborated in John C. Trautwine, *The Civil Engineer's Pocket-Book* (New York: John Wiley & Sons, 1899), 667.
11 Mainstone adds: "Because of the relatively low tensile strength, the shear strength is also low, particularly along the cleavage planes.... Compressive strength always greatly exceeds the tensile strength, usually by a factor of about ten." Mainstone, *Developments in Structural Form*, 50.
12 Carrara marble was supplied to the first Roman emperor, Augustus (63 BCE–14 CE). See Hasluck, *Building Stones*, 14.
13 As you may know, limestone is the calcareous sediment "originating from chemical action or from the breaking up of shells, corals, and the remains of other marine animals." Harry Parker, Charles Merrick Gay, and John W. MacGuire, *Materials and Methods of Architectural Construction* (1932; repr., New York: John Wiley and Sons, 1958), 113–14. Even more poetic: "A common mode of petrogenesis (creation of stone) unfolds when tiny ocean dwellers settle in their mortuary billions to the subsea muck. Limestone is a thick cemetery of mineral that had become animal now become rock again." Jeffrey Jerome Cohen, *Stone: An Ecology of the Inhuman* (Minneapolis: University of Minnesota Press, 2015), 20.
14 Williamson, *Inigo's Stones*, 203, citing a letter from "Stone the Younger," 1646–47 CE.
15 Mitchell and Mitchell, *Building Construction, Part 2*, 75.
16 Coulton, *Medieval Faith and Symbolism*, 123.
17 Rosenberg, "The Functional Aspect of the Gothic Style, Part II," 369.
18 George Wald, foreword to George Nakashima, *The Soul of a Tree: A Woodworker's Reflections* (New York: Kodansha International, 1981), xxi. Although often thought to be self-evident, this was not always well understood. Wood beams supporting the roofs in Greek temples were often set in the "wrong" or weak direction. Fowler, *A Handbook of Greek Archaeology*, 98.
19 Even when an individual stone in a wall crumbles, it tends to stay in position, since the weight of the stones above will typically hold the weakened material in place through the intensity of compression. A similar phenomenon occurs when the strength of mortar dissipates over time. As long as the wind does not erode the sand, even a brick chimney whose mortar has lost all strength will continue to stand for many decades.
20 Hasluck, *Building Stones*, 31.
21 The letter is from the architect Patrick Keely to Bishop McClusky. I speculate that the master mason raised the issue with the architect, who passed the concerns on to the client. How the fateful decision to proceed anyway was reached we will never know. The case shows the declining authority of the master mason in the nineteenth century, since a mason possessing the skill to create this fine building would never have wasted their efforts by "face bedding" the material had they full control over the purchase and quarrying of the building stone.
22 Sara E. Wermiel, "Norcross, Fuller, and the Rise of the General Contractor in the United States in the Nineteenth Century," *Proceedings of the Second International Congress on Construction History* (Exeter, UK: Short

Run Press, 2006), 3:3303, citing James O'Gorman, "O. W. Norcross, Richardson's 'Master Builder': A Preliminary Report," *Journal of the Society of Architectural Historians* 32 (1973): 108, https://www.arct.cam.ac.uk/system/files/documents/vol-3-3297-3314-wermiel.pdf.

23 Williamson, *Inigo's Stones*, 17, 38. Vitruvius's book remains widely influential today, although this particular insight of Inigo's was likely extracted from the accompanying commentary to the Italian edition by Daniele Barbaro. Williamson has also unearthed an obscure letter from the son of master mason Nicholas Stone, who executed many of Jones's famous projects, including Banqueting House. Written in the winter of 1646–47 and signed "Stone the Younger," he writes of the requirement to lay the stones "in their naturall bedd that is as thay lay in the rocke" (203).

24 "Many limestones, both travertine and Vicenza Stone being examples, must be seasoned after quarrying to be effectively worked." Peter Rockwell, *The Art of Stoneworking: A Reference Guide* (New York: Cambridge University Press, 1995), 18. Also, "All stone is better for being exposed to the air, after quarrying and before it is set, to evaporate the contained water known as quarry sap." Parker, Gay, and MacGuire, *Materials and Methods of Architectural Construction*, 127–28.

25 Using modern sawing technology, such dried-out stone blocks can still be cut and used for slabs, although sometimes this will create a visible crazing (a web of small surface cracks), particularly in larger-grained granites.

26 Parker, Gay, and MacGuire, *Materials and Methods of Architectural Construction*, 127–28. Also, "Stones were usually cut before the groundwater in them, called quarry sap, had dried out. In this condition they were softer and easier to work with a chisel or other tool." Harley J. McKee, *Introduction to Early American Masonry* (Washington, DC: National Trust for Historic Preservation, 1973), 29.

27 Hasluck, *Building Stones*, 21. Hasluck goes on to offer this exception for Bath limestone: "Stone intended for sculpturing may be removed in the green state because it is much easier to cut in that condition" (21). Logan estimates that 5 percent of the weight of the quarried block will be lost due to the curing of the stone, "a process that makes [it] tougher and more durable." Logan, *Dirt*, 99. Additionally: "It is important that the sap should be expelled before the stone is placed in a building, because when it is fixed it cannot dry out so quickly." Mitchell and Mitchell, *Building Construction, Part 2*, 77.

28 Parker, Gay, and MacGuire, *Materials and Methods of Architectural Construction*, 127. Also, Mitchell and Mitchell, *Building Construction, Part 2*, 77.

29 McPhee, *Basin and Range*, 7.

30 These sandstones are "largely derived from the primary ones by disintegration or decomposition followed by deposition, either on land or under water, and subsequent reconsolidation." Mainstone, *Developments in Structural Form*, 50.

31 Hasluck, *Building Stones*, 15.

32 Dillard, *For the Time Being*, 98, citing James S. Trefil, *A Scientist at the Seashore* (New York: Charles Scribner's Sons, 1984), 131.

33 Dillard, *For the Time Being*, 68.

34 Schwaller de Lubicz and Lamy, *The Temple in Man*, 80.

35 Hasluck, *Building Stones*, 29.

36 4,000 psi is the typical minimum threshold for compressive strength in building stone, per ASTM C170/ C170M-15b, Standard Test Method for Compressive Strength of Dimension Stone.

37 Mitchell and Mitchell, *Building Construction, Part 2*, 74–77. Although typically excluded, slate is also a viable building stone.

38 Vitruvius, *De Architectura* 2.7.5, 1688, trans. Morris Hicky Morgan, *The Ten Books on Architecture*, "Stone" (1914; repr., New York: Dover, 1960), 50. The Latin *caementiciis structuris* translates as "structural concrete," which in this context would obviously be Roman.

39 Cowan, *The Master Builders*, 64.

40 Cowan, *The Master Builders*, 205.

41 McKee, *Introduction to Early American Masonry*, 37.

42 Georges Pédro, "Clay Minerals in Weathered Rock Materials and in Soils," in *Soils and Sediments: Mineralogy and Geochemistry*, ed. Helene Parquet and Norbert Clauer (Berlin: Springer-Verlag, 1997), 1–20.

43 Michael F. Lynch, "Historic Perspectives I: The Stone Choices of Architects and Engineers, 17th Century to Contemporary Times" (lecture presented at "Building with Stone: Granite and Marble for Architectural Exteriors and Monuments," MIT, Cambridge, MA, 2004).

44 Dorothy Richter, PG, geologist, personal conversation, May 2004.

45 Patrick McDowell, "Cathedral Painted by Monet to Be Restored after Acid-Rain Damage," *Los Angeles Times*, February 3, 1991, https://www.latimes.com/archives/la-xpm-1991-02-03-mn-831-story.html.

46 Logan, *Dirt*, 121–22. For the innovative cure using paraffin and a detailed history, see Julien, "The Misfortunes of an Obelisk," 112–13, 132–33.

47 Walter R. Jaggard and Francis E. Drury, *Architectural Building Construction: A Textbook for the Architectural and Building Student* (1922; repr., Cambridge, UK: Cambridge University Press, 1938), 1:94. See also Davey, *A History of Building Materials*, 16.

48 "Building fabric" refers to the structure's exterior facing. However, for the nonstructural reality of our current method, I more often use the term "building veneer," as explained in chapter 1.

49 Alain de Botton, *The Architecture of Happiness* (New York: Pantheon Books, 2006), 97–98.

50 Flat brick jack arches are often installed with a slight upward curve to combat the illusion that the arch is sagging. A brick jack arch spanning a window opening thirty inches wide might have a camber or upward "spring" as much as three-eighths of an inch to combat the illusion.

51 Stone "floating above an opening" is often accomplished by effectively "gluing" the stone to the substrate with enhanced mortars or by supporting the floating stones with an angle iron support. Neither option is visually satisfying.

52 Jaggard and Drury, *Architectural Building Construction*, 1:94; Davey, *A History of Building Materials*, 16.

53 Mainstone, *Developments in Structural Form*, 50–51.

54 Mainstone, *Developments in Structural Form*, 38. Also, "True arch action occurs if the weight of the material produces frictional forces between layers of blocks or bricks strong enough to stop the movement of the horizontal layers in relation to each other." Cowan, *The Master Builders*, 71.

55 Michael J. Mills, "Campus Complexities: Restoration Conservation of Marble Buildings" (lecture presented at "Building with Stone: Granite and Marble for Architectural Exteriors and Monuments," MIT, Cambridge, MA, 2004).

56 Of course, this was before modern industrial diamonds made it possible to add a polish to almost anything.

57 Donald G. Presa and Jay Shockley, "Cathedral Church of Saint John the Divine and the Cathedral Close," New York City Landmarks Preservation Commission, February 21, 2017, http://s-media.nyc.gov/agencies/lpc/lp/2585.pdf, 10.

58 George W. Wickersham II, *The Cathedral Church of Saint John*, 7th ed. (Charlotte, NC: C. Harrison Conroy, n.d.), 12–14.

59 The largest completed obelisk of a single monolith is the Lateran Obelisk in Rome (105 1/2 feet, 455 tons). However, the unfinished obelisk in the Aswan quarries would have been one-third larger (137 feet), weighing in at an astonishing 1,168 tons. Kathryn Bard, *Encyclopedia of the Archaeology of Ancient Egypt* (London: Routledge, 1999), 587; Engelbach, *The Problem of the Obelisks*, 25. Also, although controversial, Lee has well-researched ideas on how the obelisks may have been freed from the living rock using levers (with the pit as the fulcrum) rather than wooden wedges. Lee, *Ancient Moonshots*, 139–54.

60 Julien, "The Misfortunes of an Obelisk," 86–87.

61 Viollet-le-Duc, *Discourses on Architecture*, 1:249.

62 All of the wonders of the ancient world were renowned for their great height: the Great Pyramid of Cheops [Giza] (the only one extant), the Hanging Gardens of Babylon, the Mausoleum at Halicarnassus, the Colossus of Rhodes, the Temple of Artemis at Ephesus, the Lighthouse at Alexandria, and the Statue of Zeus at Olympia (a four-story sculpture made of ivory and gold).
63 Rosenberg, "The Functional Aspect of the Gothic Style, Part I," 290.
64 Fred T. Hodgson, *Cyclopedia of Bricklaying, Stone Masonry, Concretes, Stuccos and Plasters* (1907; repr., Chicago: Frederick J. Drake, 1917), 73. In English brickwork common in garden walls, lacing courses occur as often as every third course, further increasing the strength of the assembly.
65 Joseph Henrich, *The WEIRDest People in the World: How the West Became Psychologically Peculiar and Particularly Prosperous* (New York: Farrar, Straus and Giroux, 2020), 103.
66 Lethaby, *Architecture*, 34.
67 Panofsky, *Gothic Architecture and Scholasticism*, 64.
68 Lethaby, *Architecture*, 65.
69 US Department of the Interior, *Riparian Area Management: Grazing Management Processes and Strategies for Riparian-Wetland Areas*, Technical Reference 1737–20, BLM/ST/ST-06/002+1737 (Denver: Bureau of Land Management, National Science and Technology Center, 2006), 105.
70 As discussed in chapter 1, sharper angles in stone create visual anachronisms, since often the brittleness of the material precludes creating too sharp a point without breaking it.
71 Koichi Barrish, chief priest, Tsubaki Grand Shrine of America, personal conversation, June 2006.
72 Lethaby, *Architecture*, 41–42.
73 E. Baldwin Smith estimated the cost as the equivalent of US $1.5 million at the time of his writing in 1938. E. Baldwin Smith, *Egyptian Architecture as Cultural Expression* (New York: D. Appleton-Century, 1938), 97.
74 Edwards, *The Pyramids of Egypt*, 285. For a small-scale contemporary example, we look to the Morgan Library in New York, designed by McKim, Mead & White and finished in 1906. The facade, constructed of white Vermont marble, achieved "zero tolerance" joints. Such "perfect" joints (without spaces) were set dry or with lime mortar. Ford, *The Details of Modern Architecture*, 1:57.
75 Mainstone, *Developments in Structural Form*, 52.
76 Harold G. Leask, *Irish Churches and Monastic Buildings* (1955; repr., Dundalk, Ireland: Litt. D. Dundalgan Press [W. Tempest)], 1987), 1:55.
77 Dennis Ogburn, "Power in Stone: The Long-Distance Movement of Building Blocks in the Inca Empire," *Ethnohistory* 51, no. 1 (2004): 106–26.
78 Lethaby coined the term "monolithism." Lethaby, *Architecture*, 34.
79 Carl G. Jung, M.-L. von Franz, Joseph L. Henderson, Jolande Jacobi, Aniela Jaffé, *Man and His Symbols* (New York: Doubleday, 1964), 209.

Chapter Four

1 Edward Tufte, personal conversation, September 2006.
2 Coldstream explains fabric rolls as "rolls of parchment membranes, about 20 cm (8 inches) wide and sewn in a continuous strip. The accounts (the fabric rolls) were kept in four-yearly terms, drawn up weekly, with receipts and expenditure separated, wages and materials differentiated and specified." Coldstream, *Medieval Craftsmen*, 21.
3 William Little, H. W. Fowler, and Jessie Coulson, *The Shorter Oxford English Dictionary on Historic Principles* (Oxford: Clarendon Press, 1933), 1:201.
4 Like many tradespeople, masons are not afraid to engage in magical thinking. It is not uncommon to hear a commanding order to an inexperienced laborer: "Bring me the brick-stretcher out of the truck! It is located next to the box of curved strings, next to the sky hook."
5 Davey, *A History of Building Materials*, 8; Mainstone, *Developments in Structural Form*, 139.
6 Lethaby quotes Petrie, specifically citing the pyramid at Meidum (he uses the antiquated spelling "Meydum"): Lethaby, *Architecture*, 60–61. See also McKee, *Introduction to Early American Masonry*, 21.
7 Lethaby, *Architecture*, 90.
8 Mainstone, *Developments in Structural Form*, 52.
9 Charles Fredrick Mitchell and George Arthur Mitchell, *Building Construction and Drawing*, 21st ed. (London: B. T. Batsford, 1953), 145; Patrick McAfee, *Stone Buildings: Conservation, Repair, Building* (Dublin: O'Brien Press, 2001), 213.
10 A. W. Lawrence, *Greek Architecture* (Harmondsworth, Middlesex, UK: Penguin Books, 1957; repr., 1962), 226.
11 Ogburn, "Power in Stone," 122.
12 Marie Jackson, et al., "Geological Basis of Vitruvius's Empirical Observations of Material Characteristics of Rock Utilized in Roman Masonry," *Proceedings of the Second International Congress of Construction History*, vol. 2 (London: Construction History Society, 2006). Jackson translates Vitruvius, *De Architectura* 2.1.1–9.
13 Victor W. von Hagen explains dramatically: "The underside of [each stone] was shaped like a diamond, truncated so as to be set into a yielding bed of gravel-sand. Each polygonal shaped stone, though massive and heavy, was cut to a jeweler's like precision, set and interlocked so that the road surface would be as solid and unyielding as Roman virtue." Victor W. von Hagen, *The Roads That Led to Rome* (Cleveland: World Publishing, 1967), 8.
14 William Manchester, *A World Lit Only by Fire: The Medieval Mind and the Renaissance* (Boston: Little, Brown and Company), 5. Williamson, describing Inigo Jones's passage from Rome to Naples via the Appian Way (built in 312 BCE), adds, "Nearly two millennia later its tough paving stones still provided a finer roadway than anything the princes of Renaissance Europe had built." Williamson, *Inigo's Stones*, 62.
15 The lessons apply to the larger building as well. Coldstream adds, "Stone is evidently a stable material. In any realistic context it will not crush under its own weight, and after an initial period of settlement during which the parts of the building find equilibrium, and providing there is no external interference such as an earthquake or a drop in the water-table (which will affect the foundations), the building will remain in a constant state." Coldstream, *Medieval Craftsmen*, 57.
16 Robert A. Johnson, *Ecstasy: Understanding the Psychology of Joy* (San Francisco: Harper & Row, 1987), 4.
17 Sabbie A. Miller et al., "Readily Implementable Techniques Can Cut CO2 Emissions from the Production of Concrete by Over 20%," *Environmental Research Letters* 11, no. 7 (2016), https://iopscience.iop.org/article/10.1088/1748-9326/11/7/074029. Portland cement, the primary ingredient in concrete, is created via a method similar to the manufacture of lime (a longer explanation of lime comes later). The calcining of the limestone to make the "clinker" for Portland cement and its subsequent crushing require vast amounts of energy (an estimated 40 percent of the total). Further, as part of the inherent chemical process, the burning of the stone itself releases vast amounts of carbon dioxide—50 percent of the total. *The Cement Sustainability Initiative: Our Agenda for Action*, pamphlet (Switzerland: World Business Council for Sustainable Development, 2002), 20, https://www.dcement.com/Upload Files/CSI_UploadFiles_7000/201108/20110831163446O3.pdf.
18 Cowan, *The Master Builders*, 256–61; "Eddystone History," Trinity House website, https://web.archive.org/web/20141108082624/http://www.trinityhouse.co.uk:80/lighthouses/lighthouse_list/eddystone.html.
19 Mainstone, *Developments in Structural Form*, 50.
20 Lawrence, *Greek Architecture*, 230; McAfee, *Stone Buildings*, 139.
21 Edwards, *The Pyramids of Egypt*, 132.
22 McAfee, *Stone Buildings*, 143.
23 US Environmental Protection Agency, "Cool Pavements," in *Reducing Urban Heat Islands: Compendium of Strategies*, https://www.epa.gov/sites/default/files/2017-05/documents/reducing_urban_heat_islands_ch_5.pdf, p. 6.

24 The ability to reflect temperature depends on the materials' solar reflective index (SRI). Natural Stone Council, *Case Study: Natural Stone Solar Reflectance Index and the Urban Heat Island Effect*, 2009, https://www.naturalstoneinstitute.org/default/assets/File/consumers/CaseStudy4_SolarReflectanceOfStone.pdf, p. 2.
25 Care should be taken in choosing limestones around pools or spas, since dark limestones will often discolor when "burned" by chlorinated water. Even saltwater systems create chlorine (acid) that will invariably react with the alkaline limestone.
26 "Self-healing in lime mortars consists of a process of dissolution, transport and re-precipitation of calcium compounds to heal cracks and fissures." Barbara Lubelli, T. G. Nijland, and R. P. J. Van Hees, "Self-Healing of Lime-Based Mortars: Microscopy Observations on Case Studies," *Heron* 56, no. 1 (2011): 1, 2, 90.
27 As witnessed by the author, April 2016.
28 Marilia Albanese, *Angkor: Splendors of the Khmer Civilization* (Bangkok: Asia Book Co., 2006),144.
29 Vincent R. Lee, personal correspondence, March 2016.
30 Stephen Boyle related this to me after touring the National Cathedral site with master mason James Bambridge in 1983. It is not known whether the National Cathedral used lime mortar/putty or the more common US standard for limestone mortar known as 6:1:1 (six parts sand, one part lime, one part Portland cement). Personal correspondence, October 2020. Also, lead is water soluble, particularly in soft, slightly acidic water. Increasingly, the use of lead for flashings is restricted by local building codes.
31 Mainstone, *Developments in Structural Form*, 53–54.
32 John Smiles et al., *The Story of John Smeaton and the Eddystone Lighthouse* (London: T. Nelson and Sons, 1882), 12–13.
33 Lethaby, *Architecture*, 43.
34 Mainstone, *Developments in Structural Form*, 185; Lethaby, *Architecture*, 59.
35 Albanese reports that the Khmer also used the metal cramp extensively. Albanese, *Angkor*, 144.
36 Dinsmoor, *The Architecture of Ancient Greece*, 174–75, 192, 235; Mainstone, *Developments in Structural Form*, 52.
37 Hislop, *Medieval Masons*, 10.
38 There is a subtle difference between liquid adhesion (in which the water molecules are less attracted to themselves and more attracted to the underside of the stone cap) and capillary action, which might be drawing or wicking surface groundwater into a stone wall at its base. Both can structurally damage masonry, but in fundamentally different ways.

Chapter Five

1 Rainer Maria Rilke, *The Letter from the Young Worker*, trans. Charlie Louth (New York: Penguin Books, 2013).
2 Ann Nottingham Kelsall, "China," in *Builders of the Ancient World: Marvels of Engineering*, Ron Fisher, et al. (Washington, DC: National Geographic Society, 1986), 170.
3 Jeanna Bryner, "Why the China Quake Was So Devastating," *Yahoo! News*, May 15, 2008, https://web.archive.org/web/20080517233751/http://news.yahoo.com/s/livescience/whythechinaquakewassodevastating.
4 Matthew 7:25.
5 Logan, *Dirt*, 104.
6 Presa and Shockley, "Cathedral Church of Saint John the Divine and the Cathedral Close," 9.
7 Stephan Boyle, personal conversation, September 2006.
8 Logan, *Dirt*, 95–96. Technically, regolith excludes soil or organic material, but in the deep view of geology, this minute layer of soil is what Logan calls "less than foam on the waves."
9 Logan, *Dirt*, 105.
10 Hodges, *Technology in the Ancient World*, 129.
11 Mainstone, *Developments in Structural Form*, 175. See also Derek Phillips, *Excavations at York Minster* (London: Her Majesty's Stationery Office, 1985), 2:64–65.
12 Saint Isaac's museum, plaque describing construction.
13 Mainstone, *Developments in Structural Form*, 175.
14 Cowan, *The Master Builders*, 110.
15 Warland, *Modern Practical Masonry*, 17.
16 Warland, *Modern Practical Masonry*, 17.
17 I paraphrase from the Italian and have recorded his words as faithfully as I can recall them.
18 Albrecht Dürer, *Rules of Proportion (Vier Büher von menschlicher Proportion)*, as related in Gerhard Rosenberg, "Functional Aspect of Gothic Style, Part II," *Journal of the Royal Institute of British Architects*, no. 43 (February 8, 1936): 366.
19 Viollet-le-Duc, *Discourses on Architecture*, 1:289–92.
20 Dinsmoor, *The Architecture of Ancient Greece*, 164, quoting Percy Gardner, *A Grammar of Greek Art* (New York: Macmillan, 1905), 39.
21 Dinsmoor, *The Architecture of Ancient Greece*, 164–65.
22 Viollet-le-Duc, *Discourses on Architecture*, 1:284.
23 Viollet-le-Duc, *Discourses on Architecture*, 1:263.
24 Viollet-le-Duc, *Discourses on Architecture*, 1:346.

Chapter Six

1 Thomas Moore, *Soul Mates: Honoring the Mysteries of Love and Relationship* (New York: HarperCollins, 1994), 121.
2 Schwaller de Lubicz and Lamy, *The Temple in Man*, 46.
3 James Hollis, *The Place of Myth in Modern Life* (Toronto: Inner City Books, 1995), 12.
4 Richard Sennett offers, "Another explanation of modern waste is that consumers are more aroused by anticipation than operation; getting the latest thing is more important than them making durable use of it.... Being able so easily to dispose of things desensitizes us to the actual objects we hold in hand." Richard Sennett, *The Craftsman* (New Haven, CT: Yale University Press, 2008), 110.
5 Lethaby, *Architecture*, 318.
6 Mainstone, *Developments in Structural Form*, 64.
7 Lars Spuybroek, *The Sympathy of Things* (London: Bloomsbury Academic, 2017), 196.
8 Nell E. Johnson, ed., *Light Is the Theme: Louis I. Kahn and the Kimbell Art Museum* (Fort Worth, TX: Kimbell Art Foundation, 1975), 53. For a cogent and related explanation of the architect Renzo Piano's method see Edward Robbins, *Why Architects Draw* (Cambridge, MA: MIT Press, 1994), 126.
9 Sennett, *The Craftsman*, 271, citing Erik Erikson, *Toys and Reasons: Stages in the Ritualization of Experience* (New York: W. W. Norton, 1977).
10 Sennett, *The Craftsman*, 295.
11 John Herman Randall Jr., "The Place of Leonardo da Vinci in the Emergence of Modern Science," *Journal of the History of Ideas* 14, no. 2 (1953): 191–202.
12 Pamela O. Long, *Artisan Practitioners and the Rise of the New Sciences, 1400–1600* (Corvallis: Oregon State University Press, 2011), 63.
13 Neil Rippingale, personal conversation, November 2004.
14 John Lobell, *Between Silence and Light: Spirit in the Architecture of Louis Kahn* (Boston: Shambhala, 1985), 40. One serious answer: brick, being absorbent, always wants to be laid into a wall rather than set as a pavement, where it will inevitably absorb water and slowly decompose. Of course, Kahn insisted, in a rhetorical flourish more dramatic than my own practical answer, the brick "likes an arch."
15 Modern science now estimates that only one acorn in ten thousand fulfills its destiny and becomes an oak tree.
16 Nakashima, *The Soul of a Tree*, 93.
17 Roger Cook, *The Tree of Life: Symbol of the Centre* (London: Thames & Hudson, 1974), 8.
18 Spuybroek, *The Sympathy of Things*, 160.
19 Gehry infamously included both asphalt and chain-link fence in his influential home built in 1978 in Santa Monica, California. It has since been suggested that these "finishes"

were, in fact, his artistic statement in reaction to the fake postmodern finishes of the time.

20 Adolf Loos, *Ornament and Crime*, trans. Shaun Whiteside (1908; repr., London: Penguin Random House UK, 2019), 185–202. Ford refers to the period in which Loos was writing as "the era of revolutionary dogma." Ford, *The Details of Modern Architecture*, 1:5.

21 "Why Is the Modern World So Ugly?," School of Life, 2020, https://www.theschooloflife.com/thebookoflife/why-is-the-modern-world-so-ugly/.

22 David Brooks, "The Moral Bucket List," *New York Times*, April 11, 2015, https://www.nytimes.com/2015/04/12/opinion/sunday/david-brooks-the-moral-bucket-list.html.

23 The US architect Fred Atherton questions whether these are actually "polar extremes or merely progression: both are load-bearing masonry, both informed by the Sacred Rules *and* profound religious conviction." Personal correspondence, October 2020.

24 De Botton, *The Architecture of Happiness*, 96.

25 Panofsky, *Gothic Architecture and Scholasticism*, 59–60.

26 Sennett, *The Craftsman*, 132. Sennett's footnote references Frank E. Brown, *Roman Architecture* (New York: G. Braziller, 1981).

27 Viollet-le-Duc, *Discourses on Architecture*, 1:375, 1:281.

28 Rosenberg, "Functional Aspect of Gothic Style, Part II," 283.

29 In fact, Panofsky points us to the flying buttresses at Caen, France, and Durham, Great Britain, which are hidden by the roofs. Panofsky, *Gothic Architecture and Scholasticism*, 57. The British architect Christopher Wren concealed the flying buttresses of Saint Paul's Cathedral in London behind a screen.

30 Viollet-le-Duc, *Discourses on Architecture*, 1:375.

31 "Tiny modules" is an exaggeration. I am referring to the generally under-scaled patterns used in faux stone patterns, not mosaic.

32 Lewis Mumford, *The South in Architecture*, Dancy Lectures, Alabama College (New York: Harcourt, Brace, 1941), 17–18.

33 Hollis, *The Place of Myth in Modern Life*, 18, citing T. S. Eliot, "Tradition and the Individual Talent," in *Critical Theory since Plato*, ed. Hazard Adams (New York: Harcourt Brace Javanovich, 1921; repr., 1971), 78.

34 William Faulkner, *Requiem for a Nun*, act 1, scene 3 (New York: Random House, 1951).

35 See Cowan, *The Master Builders*, 163, which points to Walter Gropius, echoing Friedrich Nietzsche.

36 Lawrence, *Greek Architecture*, 233, citing Aristotle, *Politics*, book 7, chapter 2.

37 Lethaby, *Architecture*, 10.

38 Lethaby, *Architecture*, 129.

39 James Hollis, *The Eden Project: In Search of the Magical Other* (Toronto: Inner City Books, 1998), 113.

40 T. S. Eliot, "Burnt Norton," in *Four Quartets* (New York: Harcourt Brace, 1943).

41 Hollis, *The Eden Project*, 115.

42 "Quantum physics shows us "the *dancing universe*; the ceaseless flow of energy going through an infinite variety of patterns. This is the Dionysian energy, the dance of the Maenads, the power of life that flows through all of us and unites us with heaven and earth." Johnson, *Ecstasy*, 11, quoting Fritjof Capra, *The Tao of Physics* (Boulder, CO: Shambhala, 1975), 276.

43 See Thomas Moore, *The Soul of Sex: Cultivating Life as an Act of Love* (New York: HarperCollins, 1998), 88.

44 Hollis, *The Eden Project*, 125–26.

45 "Accept this then for a universal law, that neither architecture nor any other noble work of man can be good unless it be imperfect." John Ruskin, *Stones of Venice* (1853; repr., New York: Peter Fenelon Collier & Son, 1900), 2:172. Ford points us to Ruskin's clarification, "Accurately speaking, no good work whatever can be perfect, and the demand for perfection is always a sign of a misunderstanding of the ends of art." Ford, *The Details of Modern Architecture*, 1:9.

46 Students of the philosopher Martin Heidegger will notice a parallel with his concept of "equipmental totality," where "a broken hammer will 'light up' the workshop." Martin Heidegger, "The Worldhood of the World," section 16, in *Being and Time*, trans. John Macquarrie (1927; repr., London: SMC Press, 1962), 97.

47 Spuybroek, *The Sympathy of Things*, 197.

48 Sennett, *The Craftsman*, 84, 81, 109.

49 Nakashima, *The Soul of a Tree*, 125.

50 Ford, *The Details of Modern Architecture*, 1:7–9, quotes Harry Stuart Goodhart-Rendel: "That the marks of the worker's tool can add positive value to the finished work, a tenet that is less based upon aesthetics than upon sociological...was the foundation for the whole movement known as the Arts and Crafts." Harry Stuart Goodhart-Rendel, *English Architecture since the Regency: An Interpretation* (St. Claire, MI: Scholarly Press, 1953; repr., 1977), 192.

51 Spuybroek, *The Sympathy of Things*, 199.

52 Jun'ichirō Tanizaki, *In Praise of Shadows* (Stony Creek, CT: Leete's Island Books, 1977), 11.

53 The parallels with the past may be stronger than we realize. We might imagine that in the small towns (by today's standards) where most people lived, everyone was socially connected, but Long points out, "Shoemakers and university professors still lived and worked worlds apart in the late sixteenth century, as they had in the twelfth." Long, *Artisan Practitioners and the Rise of the New Sciences, 1400–1600*, 128.

54 Amy Gansell, private conversation and correspondence, 2004.

55 K. Michael Hays, *Sanctuaries: Last Works of John Hejduk* (New York: Whitney Museum of American Art, 2002), book jacket.

56 De Botton, *The Architecture of Happiness*, 81.

57 "The less we understand of what our fathers and forefathers sought, the less we understand ourselves, and thus we help with all our might to rob the individual of his roots and his guiding instincts." Carl Jung, *Memories, Dreams, Reflections* (London: Collins and Routledge & Kegan Paul, 1963), 223.

58 "We count for something only because of the essential we embody, and if we do not embody that life is wasted." Jung, *Memories, Dreams, Reflections*, 300.

59 "Such views ignore the real achievements of our ancestors and constitute the ultimate racism: they belittle the abilities and ingenuity of the human species as a whole." John Flenley and Pau Bahn, *The Enigmas of Easter Island, Island on the Edge* (Oxford, UK: Oxford University Press, 2002), 114.

60 Jung, *Memories, Dreams, Reflections*, 280.

61 James Hollis, personal conversation, 2017. Hollis echoes Jung, who reportedly said, "We all walk in shoes too small."

Bibliography

Ackerman, James S. 1949. "*Ars sine scientia nihil est*, Gothic Theory of Architecture at the Cathedral of Milan." *Art Bulletin* 31:84–111.

Acocella, Alfonso. 2006. *Stone Architecture: Ancient and Modern Construction Skills*. Lucca, Italy: Lucense SCpA.

Albanese, Marilia. 2006. *Angkor: Splendors of the Khmer Civilization*. Bangkok: Asia Book Co.

Andrews, Francis B. (1925) 1993. *The Mediaeval Builder and His Methods*. New York: Barnes and Noble.

Bard, Kathryn. 1999. *Encyclopedia of the Archaeology of Ancient Egypt*. London: Routledge.

Brooks, David. 2015. "The Moral Bucket List." *New York Times*, April 11, 2015, https://www.nytimes.com/2015/04/12/opinion/sunday/david-brooks-the-moral-bucket-list.html.

Brunés, Tons. 1967. *The Secrets of Ancient Geometry and Its Use*. 2 vols. Copenhagen: International Science Publishers.

Campbell, James W. P. 2004. *Brick: A World History*. London: Thames & Hudson.

Coldstream, Nicola. 1991. *Medieval Craftsmen: Masons and Sculptors*. Toronto: University of Toronto Press.

Cook, Roger. 1974. *The Tree of Life: Symbol of the Centre*. London: Thames & Hudson.

Cowan, Henry J. 1977. *The Master Builders: A History of Environmental Design from Ancient Egypt to the Nineteenth Century*. New York: John Wiley & Sons.

Cranage, David Herbert Somerset. 1951. *Cathedrals and How They Were Built*. Cambridge, UK: Cambridge University Press.

Crawford, Harriet. (1991) 2000. *Sumer and the Sumerians*. Cambridge, UK: Cambridge University Press.

Critchlow, Keith. 1979. *Time Stands Still: A New Light on Megalith Science*. London: Gordon Fraser.

Curran, Brian A., Anthony Grafton, Pamela O. Long, and Benjamin Weiss. 2009. *Obelisk: A History*. Cambridge, MA: MIT Press.

Dale, T. Nelson. 1908. *The Chief Commercial Granites of Massachusetts, New Hampshire and Rhode Island*. Washington, DC: Government Printing Office.

De Botton, Alain. 2006. *The Architecture of Happiness*. New York: Pantheon Books.

De Honnecourt, Villard. 1959. *The Sketchbook of Villard de Honnecourt*. Bloomington: Indiana University Press.

Dillard, Annie. 1999. *For the Time Being*. New York: Knopf.

Dinsmoor, William Bell. (1909) 1950. *Architecture of Ancient Greece: An Account of Its Historic Development*. London: B. T. Batsford.

Domenico, Fontana. (1590) 2002. *Della trasportatione dell'obelisco vaticano*. Oakland, CA: Octavo.

Edwards, I. E. S. (1947) 1972. *The Pyramids of Egypt*. New York: Viking Press.

Eliot, T. S. 1943. *Four Quartets*. New York: Harcourt Brace.

Engelbach, Reginald. 1923. *The Problem of the Obelisks: From a Study of the Unfinished Obelisk at Aswan*. New York: George H. Doran.

Erlade-Brandenburg, Alain. 1995. *The Cathedral Builders of the Middle Ages*. London: Thames & Hudson.

Faude, Oliver, Wilfried Kindermann, and Tim Meyer. 2009. "Lactate Threshold Concepts: How Valid Are They?" *Sports Medicine* 39 (6): 469–90.

Faulkner, William. 1951. *Requiem for a Nun*. New York: Random House.

Flenley, John, and Paul Bahn. 2002. *The Enigmas of Easter Island*. Oxford: Oxford University Press.

Fletcher, Sir Banister. (1896) 1987. *A History of Architecture*. 19th ed. London: Butterworths.

Ford, Edward R. 2003. *The Details of Modern Architecture*. Cambridge, MA: MIT Press.

Fowler, Harold North, James Rignall Wheeler, and Gorham Phillips Stevens. 1909. *A Handbook of Greek Archaeology*. New York: American Book Company.

Geikie, Archibald. (1882) 1903. *Text-Book of Geology*. New York: Macmillan.

Giedion, Sigfried. 1964. *The Eternal Present: The Beginnings of Architecture*. New York: Pantheon Books.

Gimpel, Jean. 1961. *The Cathedral Builders*. New York: Grove Press.

Giry, A. 1878. *Notice sur un traité du Moyen âge intitulé: De Colorisbus et Artibus Romanorum*. Bibliothèque de l'École des Hautes Études, fasc. 35. Paris.

Gould, Robert Freke. 1886. *The History of Freemasonry*. New York: John C. Yorston.

Harvey, John. 1971. *The Master Builders: Architecture in the Middle Ages*. New York: McGraw Hill.

Hasluck, Paul N. 1904. *Building Stones*. London: Cassell.

Hays, K. Michael. 2002. *Sanctuaries: Last Works of John Hejduk*. New York: Whitney Museum of American Art.

Heidegger, Martin. (1927) 1962. *Being and Time*. Translated by John Macquarrie. London: SMC Press.

Henrich, Joseph. 2020. *The WEIRDest People in the World: How the West Became Psychologically Peculiar and Particularly Prosperous.* New York: Farrar, Straus and Giroux.

Herbermann, Charles George. 1913. *The Catholic Encyclopedia.* New York: Robert Appleton.

Hero of Alexandria. 1966. *Mechanics.* In M. R. Cohen and I. E. Drabkin, *A Source Book in Greek Science.* Cambridge, MA: Harvard University Press.

Heyman, Jacques. 1966. "The Stone Skeleton." *International Journal of Solids and Structures* 2 (2): 249–56.

———. 1967. "Beauvais Cathedral." *Transactions of the Newcomen Society* 40 (1): 15–35.

Hislop, Malcolm. 2009. *Medieval Masons.* Oxford: Shire.

Hodges, Henry. 1970. *Technology in the Ancient World.* New York: Alfred A. Knopf.

Hodgson, Fred T. (1907) 1917. *Cyclopedia of Bricklaying, Stone Masonry, Concretes, Stuccos and Plasters.* Chicago: Frederick J. Drake.

Hollis, James. 1995. *The Place of Myth in Modern Life.* Toronto: Inner City Books.

———. 1998. *The Eden Project: In Search of the Magical Other.* Toronto: Inner City Books.

Jaggard, Walter R., and Francis E. Drury. (1922) 1938. *Architectural Building Construction: A Textbook for the Architectural and Building Student.* 4th ed. Cambridge, UK: Cambridge University Press.

Johnson, Nell E., ed. 1975. *Light Is the Theme: Louis I. Kahn and the Kimbell Art Museum.* Fort Worth, TX: Kimbell Art Foundation.

Johnson, Robert A. 1987. *Ecstasy: Understanding the Psychology of Joy.* San Francisco: Harper & Row.

Julien, Alexis A. 1893. "The Misfortunes of an Obelisk." *Journal of the American Geographical Society of New York* 25 (1): 66–137.

Jung, Carl. 1963. *Memories, Dreams, Reflections.* London: Collins and Routledge & Kegan Paul.

Jung, Carl G., M.-L. von Franz, Joseph L. Henderson, Jolande Jacobi, and Aniela Jaffé. 1964. *Man and His Symbols.* Garden City, NY: Doubleday.

Kelsall, Ann Nottingham. 1986. "China." In *Builders of the Ancient World: Marvels of Engineering*, edited by Ron Fisher et al. Washington, DC: National Geographic Society.

Knoop, Douglas. 1939. *Pure Ancient Masonry.* Private printing.

Kostof, Spiro, ed. 1977. *The Architect: Chapters in the History of the Profession.* New York: Oxford University Press.

Larsen, C. E. 1975. "The Mesopotamian Delta Region: A Reconstruction of Lees and Falcon." *Journal of the American Oriental Society* 95:43–57.

Lawlor, Robert. 1983. *Sacred Geometry: Philosophy and Practice.* London: Thames & Hudson.

Leask, Harold G. (1955) 1987. *Irish Churches and Monastic Buildings.* Dundalk, Ireland: Litt. D. Dundalgan Press (W. Tempest).

Lee, Vincent R. 2013. *Ancient Moonshots: Megalithic Mysteries from before Technology.* Jackson Hole, WY: Sixpac Manco.

Lethaby, William Richard. (1911) 1929. *Architecture: An Introduction to the History and Theory of the Art of Building.* London: Thornton Butterworth.

Little, William, H. W. Fowler, and Jessie Coulson. (1933) 1986. *The Shorter Oxford English Dictionary on Historic Principles.* Oxford: Clarendon Press.

Lobell, John. 1985. *Between Silence and Light: Spirit in the Architecture of Louis Kahn.* Boston: Shambhala.

Logan, William Bryant. 1995. *Dirt: The Ecstatic Skin of the Earth.* New York: Riverhead Books.

Long, Pamela O. 2011. *Artisan Practitioners and the Rise of the New Sciences, 1400–1600.* Corvallis: Oregon State University Press.

Loos, Adolf. (1908) 2019. *Ornament and Crime.* Translated by Shaun Whiteside. London: Penguin Random House UK.

Lstiburek, Joseph W. 2010. "Mind the Gap, Eh?" *ASHRAE Journal* 52 (1): 57–63.

Lubelli, Barbara, T. G. Nijland, and R. P. J. Van Hees. 2011. "Self-Healing of Lime-Based Mortars: Microscopy Observations on Case Studies." *Heron* 56 (1): 75–91.

Mahlstedt, Thomas F., and Lauren Cook. 2003. *Granite Railway Quarry, Phase II Stages 1 & 2*, vol. 3. Boston: Metropolitan District Commission.

Mainstone, Rowland J. 1975. *Developments in Structural Form.* Cambridge, MA: MIT Press.

Manchester, William. 1992. *A World Lit Only by Fire: The Medieval Mind and the Renaissance.* Boston: Little, Brown.

McAfee, Patrick. 2001. *Stone Buildings: Conservation, Repair, Building.* Dublin: O'Brien Press.

McDowell, Patrick. 1991. "Cathedral Painted by Monet to Be Restored after Acid-Rain Damage." *Los Angeles Times*, February 3, 1991, https://www.latimes.com/archives/la-xpm-1991-02-03-mn-831-story.html.

McKee, Harley J. 1973. *Introduction to Early American Masonry.* Washington, DC: National Trust for Historic Preservation.

McPhee, John. 1981. *Basin and Range*. New York: Farrar, Straus and Giroux.
Merrifield, Mary P., ed. (1849) 1967. *Original Treatises on the Arts of Painting*. New York: Dover.
Miller, Sabbie A., Arpad Horvath, and Paulo J. M. Monteiro. 2016. "Readily Implementable Techniques Can Cut CO2 Emissions from the Production of Concrete by Over 20%." *Environmental Research Letters* 11 (7): https://iopscience.iop.org/article/10.1088/1748-9326/11/7/074029.
Mitchell, Charles F., and A. M. Mitchell. (1893) 1947. *Building Construction, Part 2, Advanced Course*. 16th ed. London: B. T. Batsford.
Mitchell, Charles F., and George Arthur Mitchell. 1953. *Building Construction and Drawing*. 21st ed. London: B. T. Batsford.
Moore, Thomas. 1994. *Soul Mates: Honoring the Mysteries of Love and Relationship*. New York: HarperCollins.
———. 1998. *The Soul of Sex: Cultivating Life as an Act of Love*. New York: HarperCollins.
Mumford, Lewis. 1941. *The South in Architecture*. Dancy Lectures, Alabama College. New York: Harcourt, Brace.
Nair, Stella. 2015. *At Home with the Sapa Inca: Architecture, Space, and Legacy at Chinchero*. Austin: University of Texas Press.
Nakashima, George. 1981. *The Soul of a Tree: A Woodworker's Reflections*. New York: Kodansha International.
Natural Stone Council. 2009. *Case Study: Natural Stone Solar Reflectance Index and the Urban Heat Island Effect*. https://www.naturalstoneinstitute.org/default/assets/File/consumers/CaseStudy4_SolarReflectanceOfStone.pdf.
Ogburn, Dennis. 2004. "Power in Stone: The Long-Distance Movement of Building Blocks in the Inca Empire." *Ethnohistory* 51 (1): 106–26.
Panofsky, Erwin. 1948. *Abbot Suger: On the Abbey Church of St.-Denis and Its Art Treasures*. Princeton, NJ: Princeton University Press.
———. 1959. *Gothic Architecture and Scholasticism*. New York: Meridian Books.
Parker, Harry, Charles Merrick Gay, and John W. MacGuire. (1932) 1958. *Materials and Methods of Architectural Construction*. New York: John Wiley and Sons.
Parrott III, Les. 2004. *Love Talk*. Grand Rapids, MI: Zondervan.
Payne, Christopher. 2020. "Vital Vessels." *New Yorker*, December 7, 2020, 48–57.
Pédro, Georges. 1997. "Clay Minerals in Weathered Rock Materials and in Soils." In *Soils and Sediments: Mineralogy and Geochemistry*, edited by Helene Parquet and Norbert Clauer, 1–20. Berlin: Springer-Verlag.
Phillips, Derek. 1985. *Excavations at York Minster*. London: Her Majesty's Stationery Office.
Plato. *Cratylus*.
Presa, Donald G., and Jay Shockley. 2017. "Cathedral Church of Saint John the Divine and the Cathedral Close." New York City Landmarks Preservation Commission, February 21, 2017, http://s-media.nyc.gov/agencies/lpc/lp/2585.pdf.
Protzen, Jean-Pierre. 1983. "Inca Quarrying and Stonecutting." *Journal of the Society of Architectural Historians* 44 (2): 161–82.
Protzen, Jean-Pierre, and Stella Nair. 2014. *The Stones of Tiahuanaco: A Study of Architecture and Construction*. Los Angeles: Cotsen Institute of Archaeology Press.
Randall Jr., John Herman. 1953. "The Place of Leonardo da Vinci in the Emergence of Modern Science." *Journal of the History of Ideas* 14 (2): 191–202.
Ravenscroft, W. 1910. *The Comancines*. London: Elliot Stock.
Rich, Ben R., and Leo Janos. 1994. *Skunk Works*. New York: Little, Brown.
Richards, John Chatterton. 1940. "A New Manuscript of Heraclius." *Speculum* 15 (3): 225–71.
Rilke, Rainer Maria. 2013. *The Letter from the Young Worker*. Translated by Charlie Louth. New York: Penguin Books.
Ritti, Tullia, Klaus Grewe, and Paul Kessener. 2007. "A Relief of a Water-Powered Stone Saw Mill on a Sarcophagus at Hierapolis and Its Implications." *Journal of Roman Archaeology* 20:138–63.
Robbins, Edward. 1994. *Why Architects Draw*. Cambridge, MA: MIT Press.
Rockwell, Peter. 1995. *The Art of Stoneworking: A Reference Guide*. Cambridge, UK: Cambridge University Press.
Ronnberg, Ami, Kathleen Martin et al. 2010. *The Book of Symbols: Reflections on Archetypal Images*. Cologne: Taschen.
Rosenberg, Gerhard. 1936. "The Functional Aspect of the Gothic Style, Part I." *Journal of the Royal Institute of British Architects* 43:273–90.
———. 1936. "The Functional Aspect of the Gothic Style, Part II." *Journal of the Royal Institute of British Architects* 43:364–71.

Routledge, Katherine. (1919) 2007. *The Mystery of Easter Island*. New York: Cosimo Classics.
Ruskin, John. (1853) 1900. *Stones of Venice*. New York: Peter Fenelon Collier & Son.
Schwaller de Lubicz, R. A., and Lucie Lamy. (1949) 1977. *The Temple in Man: The Secrets of Ancient Egypt*. Brookline, MA: Autumn Press.
Sennett, Richard. 2008. *The Craftsman*. New Haven, CT: Yale University Press.
Smiles, John, et al. 1882. *The Story of John Smeaton and the Eddystone Lighthouse*. London: T. Nelson and Sons.
Smith, E. Baldwin. 1938. *Egyptian Architecture as Cultural Expression*. New York: D. Appleton-Century.
Snelling, Andrew A., and John Woodmorappe. 1998. "The Cooling of Thick Igneous Bodies on a Young Earth." In *Proceedings of the Fourth International Conference on Creationism*, 527–45. Pittsburgh: Creation Science Fellowship.
Spuybroek, Lars. 2017. *The Sympathy of Things*. London: Bloomsbury Academic.
Svanberg, Jan. 1983. *Master Masons*. Uppsala, Sweden: Carmina.
Tanizaki, Jun'ichirō. 1977. *In Praise of Shadows*. Stony Creek, CT: Leete's Island Books.
Trautwine, John C. 1899. *The Civil Engineer's Pocket-Book*. New York: John Wiley & Sons.
US Department of the Interior. 2006. *Riparian Area Management: Grazing Management Processes and Strategies for Riparian-Wetland Areas*. Technical Reference 1737-20, BLM/ST/ST-06/002+1737. Denver: Bureau of Land Management, National Science and Technology Center.
US Environmental Protection Agency. 2012. "Cool Pavements." In *Reducing Urban Heat Islands: Compendium of Strategies*. https://www.epa.gov/sites/default/files/2017-05/documents/reducing_urban_heat_islands_ch_5.pdf.
US Geological Survey. 2019. *Mineral Commodity Summaries*. Reston, VA: US Geological Survey.
Viollet-le-Duc, Eugène Emmanuel. (1889) 1959. *Discourses on Architecture*. Translated by Benjamin Bucknall. New York: Grove Press.
Vitruvius. (1688) 1960. *De Architectura*. In *The Ten Books on Architecture*. Translated by Morris Hicky Morgan. New York: Dover.
von Hagen, Adriana, and Craig Morris. 1998. *The Cities of the Ancient Andes*. London: Thames & Hudson.
von Hagen, Victor W. 1967. *The Roads That Led to Rome*. Cleveland: World Publishing.
Wade, Nicholas. 2005. "Those Ancient Incan Knots? Tax Accounting, Researchers Suggest." *New York Times*, August 16, 2005, https://www.nytimes.com/2005/08/16/science/those-ancient-incan-knots-tax-accounting-researchers-suggest.html.
Warland, E. G. 1929. *Modern Practical Masonry*. London: Library Press Limited.
Wickersham II, George W. n.d. *The Cathedral Church of Saint John*. 7th ed. Charlotte, NC: C. Harrison Conroy.
Williamson, Tom. 2012. *Inigo's Stones: Inigo Jones, Royal Marbles and Imperial Power*. Padstow, England: Matador.

Index

Page numbers in **bold** indicate illustrations and their captions.

Credits

31: Norman Davey, *A History of Building Materials* (London: Phoenix House, 1961), 225.

33 left: Miniature by Raban Maur, "De Originibus" (ca. 842), published in Pierre du Colombier, *Les Chantiers des Cathédrales* (1953; repr., Paris: A. & J. Picard, 1973), 8.

34 top: Paul Kessener.

37, 112, 172 top and bottom, 191 right, 194: Pamela Rhodes.

57: P. E. Newberry, from I. E. S. Edwards, *The Pyramids of Egypt* (1947; repr., New York: Viking Press, 1972), part 1, plate XV, 202.

71: Francis B. Andrews, *The Mediaeval Builder and His Methods* (1925; repr., New York: Barnes and Noble, 1993), 10.

114: Alfonso Acocella, *Stone Architecture: Ancient and Modern Construction Skills* (Lucca, Italy: Lucense SCpA, 2006), 210. Drawing is unattributed.

145: iStock by Getty Images. 2100322930.

152: Robert Valenti.

181: G. A. Breymann, et al. *Allgemeine Baukonstruktionslehre: mit besonderer Beziehung auf das Hochbauwesen; ein Handbuch zu Vorlesungen und zum Selbstunterricht*, vol. 1 (Leipzig: J. M. Gebhardt, 1896), Plate 10.

243: Nicholas Moore.

Published by
Princeton Architectural Press
A division of Chronicle Books LLC
70 West 36th Street
New York, NY 10018
papress.com

Printed and bound in China

28 27 26 25   4 3 2 1   First edition

ISBN 978-1-7972-3008-5

Library of Congress Control Number: 2024036078

Editor: Jennifer N. Thompson
Designer: Paul Wagner